1V CMOS G_m-C Filters

ANALOG CIRCUITS AND SIGNAL PROCESSING SERIES

Consulting Editor: **Mohammed Ismail.** ***Ohio State University***

For other titles published in this series, go to
www.springer.com/series/7381

Tien-Yu Lo • Chung-Chih Hung

1V CMOS G_m-C Filters

Design and Applications

Springer

Dr. Tien-Yu Lo
MediaTek Inc.
No.1, Dusing Rd. 1
Hsinchu 30078
Hsinchu Science Park
Taiwan
tienyu.lo@gmail.com

Assoc. Prof. Chung-Chih Hung
National Chiao Tung University
Department of Electrical Engineering
1001 University Road
Hsinchu 300
Taiwan
cchung@mail.nctu.edu.tw

ISBN 978-90-481-2409-1 e-ISBN 978-90-481-2410-7
DOI 10.1007/978-90-481-2410-7
Springer Dordrecht Heidelberg London New York

Library of Congress Control Number: PCN applied for

Printed on acid-free paper

Springer is part of Springer Science+Business Media (www.springer.com)

This book is dedicated to Our Families

Preface

There are growing demands for low-voltage circuits and systems, especially for system-on-a-chip applications. When switched to low supply voltage, digital circuits do not suffer the degradation in performance. On the other hand, the circuit performance of analog circuits are strongly affected by reduced supply voltage. In addition, chip area should also be taken into consideration to reduce the cost of advanced multi-function SOC design. Therefore, there is urgent need to develop new design techniques for analog circuits at 1-V supply.

In Chapter 2, novel transconductors with the applications for wireless and wireline systems are introduced. The transconductor is a basic building block for analog circuits, such as the G_m-C filter, continuous-time delta-sigma modulator, voltage-controlled oscillator, and multiplier. First, the configuration of a linearized transconductor for low-voltage and high-frequency applications is proposed. By using double pseudo-differential pairs and the source degeneration structure under nano-scale CMOS technology, the nonlinearity caused by short channel effect from small feature size can be minimized. Then, another transconductor with pseudo-differential structures is proposed. The linearity is improved by mobility compensation techniques as the device size is scaled down to achieve high-speed operation. Short channel effects in the nano-scale technology are discussed and eliminated, and the results show superior performance even at high-speed operation. Finally, a transconductor with the specific target for ADSL2+ application is discussed. A precise model is adopted to eliminate short channel effects and high linearity performance can thus be achieved.

Chapter 3 focuses on the basic concept of the G_m-C implementation. A G_m-C filter with a fifth-order Elliptic prototype and a wide tuning range for very low frequency is discussed. The transconductor can work from the weak inversion region to the strong inversion region to maximize the transconductance tuning range. The transconductance can be tuned by changing its bias current. Through the use of switching technology, the filter can be applied to biomedical systems, audio systems and part of wireless systems.

Three multi-mode channel selection filters for the Zero-IF direct conversion receiver are presented in Chapter 4. The specific transconductors with required functions are designed through the use of a third-order Butterworth prototype. The proposed circuit first uses the forward voltage follower structure and an active resistor to implement the linear transconductor. The wide tuning range would be suitable for the specifications of IEEE 802.11a/b/g Wireless LANs, Wideband CDMA, cdma2000, and Bluetooth. Then, the other transconductor is presented. The transconductor includes a voltage-to-current converter and a current multiplier. Voltage-to-current conversion employs linear-region MOS transistors and the circuit features high linearity over a wide input swing. The current multiplier which operates in the weak inversion region provides a wide transconductance tuning range without degrading the linearity. The tuning range and the linearity performance would be suitable for the wireless specifications of GSM, Bluetooth, cdma2000, and Wideband CDMA. Finally, another transconductor uses the same concept as that of the previous one. Linear-region MOS transistors are employed to perform the voltage-to-current conversion. The wide tuning range can be achieved by the current multiplier following the linear voltage-to-current converter.

Two high-speed filters with a fourth-order equiripple prototype are analyzed in Chapter 5. The high-speed filter can be used for pulse signal systems. One is designed for the hard disk storage systems, and a novel automatic tuning circuit is also implemented to account for variations in process and temperature. A high-speed transconductor with the inverter structure is realized. Transconductance tuning can be achieved by adjusting the bulk voltage using the Deep-NWELL technology. Then, a high-performance G_m-C equiripple linear phase low-pass filter for UWB wireless application is presented. The proposed transconductor is designed under low supply voltage while its gain, excess phase, and linearity are well maintained.

This book is organized based upon the Ph.D. dissertation in National Chiao Tung University, Taiwan. It includes details and measurement results for each research project, and the analyses made with circuits operating at 1-V supply. The projects were supported by the National Science Council of Taiwan, while the chips were fabricated by the National Chip Implementation Center of Taiwan.

This book would not been published without the help and kindness of many individuals. First of all, we would like to thank Prof. Mohammed Ismail of the Ohio State University for his helpful comments. He is indeed a world leader in the research and education of analog and RF circuit designs. Some sections are enlightened by his influential work. We would like to express our gratitude to Mr. Chih-Lung Kuo, who has made significant contribution to this book. He spent days and nights typesetting and editing the initial version. A special thank goes to all the members of NCTU Analog Integrated Circuit Lab for their generous assistance and insightful technical discussion over the years. Finally, the first author would like to thank Dr. Hung-Sung Li at MediaTek Inc. for his guidance, encouragement, and support.

Tien-Yu Lo
Chung-Chih Hung

Contents

List of Figures

List of Tables

Chapter 1
Motivation

1.1 Introduction

Low-voltage VLSI circuits have received significant attention in recent years. The supply voltage has decreased following the advancement of process technology, and the operation frequency of CMOS can also go up higher than 100 GHz. The demands for nano-scale applications are mainly classified into technology-driven, design-driven, and market-driven. They are mainly to reduce the minimum feature size to scale down the chip area, to fabricate millions of transistors on a single chip to save cost, and to the increase market demands for communication electronic products. These demands seem to be independent of each other. However, the advances in VLSI technology, circuit design, and product market are actually interrelated.

In the past decade, CMOS technology has played a major role in the rapid advancement and the increased integration of VLSI systems. CMOS devices feature high input impedance, extremely low offset switches, high packing density, low switching power consumption, and thus can be easily scaled. The minimum feature size of a MOS transistor has been decreasing [1–3]. Current VLSI technology is scaled down to around 90 nm. Scaling down the transistor sizes can then integrate more circuit components in a single chip, so the circuit area and thus its cost will be reduced. Besides this economic consideration, smaller geometry usually lowers the parasitic capacitance, which leads to higher operating speed. When a MOS transistor size is decreased, not only its channel length and width are reduced, but also the thickness of the gate oxide. As a MOS transistor has a thinner gate oxide, in order to prevent the transistor from breakdown due to higher electrical field across the gate oxide and to ensure its reliability, the supply voltage needs to be reduced [4–8]. Since the digital circuits are becoming more and more popular, the computer-aided design tools for digital circuits are getting matured, the digital circuits are certainly occupying most of the fabricated chip area, and the electrical characteristics of MOS transistors are getting optimized mainly for digital circuits. Switching to use nano-scale devices, digital circuits do not suffer the degradation of their performances too much.

T.-Y. Lo and C.-C. Hung, *1V CMOS* G_m-C *Filters: Design and Applications*, Analog Circuits and Signal Processing,

On the other hand, for analog circuits, the circuit performances, such as gain, dynamic range, speed, bandwidth, linearity, etc., are strongly affected by using nano-scale technology. Therefore, new design techniques for nano-scale analog circuits are required to be developed. Moreover, since we are living in an analog world, it is inevitable to use analog signal processing. Modern analog and mixed-signal VLSI applications in areas such as telecommunications, smart sensors, battery-operated consumer electronics and artificial neural computation require CMOS analog design solutions. Thus, analog signal and information processing in nano-scale technology is really a field requiring more devotion.

1.2 Applications

In the systems that interface with real word, the processed signal would be measured with unwanted noise. A filter is usually used to get rid of the unwanted noise and to reject the surrounding interface. Thus, filters are important block for specified frequency of signals and they are essential for many applications. They can be used to band-limit signals in wireline and wireless communication systems. These filters operate on continuous-time fashion because the systems interface with real analog world. Figure 1.1 shows the required low-pass filter for specified applications.

There are two kinds of the filters: digital filters and analog filters. The analog filters process the continuous data rather than the digital data for digital filters. The analog filters can be further divided into passive filters and active filters. The elements of a passive filter are passive, and a passive filter includes resistors, capacitors, inductors, and transformers. On the contrary, the active filters include active devices. A large area is required for the passive filter, and then the active filter is more suitable in CMOS technology.

Active filters can be classified into Active-RC, Switched-Capacitor, G_m-C, and LC filters. The Active-RC and Switched-Capacitor filters are only suitable for low to medium frequency applications. For high frequencies, the settling problem of

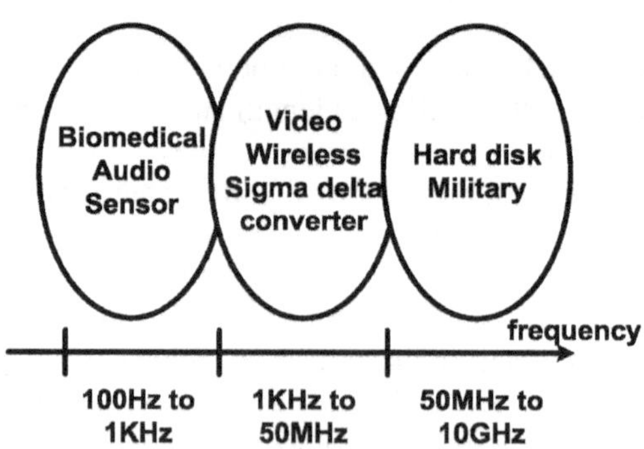

Fig. 1.1 The filter operation frequency for various applications

amplifiers would affect the filter performance since very wide bandwidth and high unity-gain frequency are hard to be achieved. For systems in the GHz range, LC filters are a better choice since the required values of L and C are small. However, Q enhancement is needed for LC filters because of low inductor quality factors. The G_m-C filters, which operate on open loop topology, would be sufficient for low to high frequency range. Thus, the G_m-C architecture can be implemented for various applications. In addition, it is noted that the performance of the G_m-C filter is highly dependent on the performance of the transconductor.

Another issue is the automatic tuning circuit. The frequency response and the quality factor should be maintained owing to process, supply voltage and temperature variations. Thus, a high performance automatic tuning circuit is required for continuous-time active filters.

1.3 Organization

This book focuses on the transconductor with the implementation of G_m-C filter for wireless and wireline applications at 1-V supply. This book is organized into four parts. First, we introduce the basic concepts of transconductors. The circuits with varies linearity architecture are discussed. Three modified transcontuctors which can operate under high frequency or high linearity are present. Next, the filter synthesis from the passive prototype to the active prototype is introduced. The non-ideality caused by the integrator is also discussed. We illustrate some important parameters to illustrate filter performance. A wide tuning range filter is presented as an example. Then, the zero-IF architecture for wireless communication is introduced. Three modified channel selection multi-mode filter for various wireless applications are presented. The channel selection filters can be suitable for the specifications of GSM, Bluetooth, cdma2000, Wideband CDMA, and IEEE 802.11 a/b/g/n Wireless LANs. At last, we introduce the equiripple filter which is required for pulse signal channel. The signal processing requires constant group delay. The basic concept of automatic tuning circuit is also introduced. A high-speed filter used for hard disk read channel with a modified frequency tuning circuit is presented. Finally, another filter used for UWB application is presented.

Chapter 2
Transconductor

2.1 Introduction

This chapter shows the basic concept and improved design of a transconductor. The transconductor is one of the most important building blocks in analog and mix-mode circuits, including multipliers [9, 10], continuous-time G_m-C filters [11, 12], voltage controlled oscillators [13], and continuous-time sigma-delta modulators [14]. Its main idea is to convert the input voltage into the output current with a linear transformation factor. The active device is used for replacing passive devices owing to power and area consideration with the tradeoff of the non-ideal performance. The main non-ideal characteristics of the transconductor are the limited linear input range, limited output impedance, finite signal-to-noise ratio, and finite bandwidth. The linear performance is the most important issue in the transconductor design. Moreover, as the feature size of CMOS technology scales down with supply voltage, the dynamic range, bandwidth, and power consumption will be limited under specific linearity.

2.2 Review of CMOS Transconductors

The transconductor in CMOS process is required for the system-on-a-chip strategy. Thus, the following section discusses the reported basic linearity technique in CMOS process.

2.2.1 The Source Degenerated Transconductor

Figure 2.1 shows circuit implementation of the source degeneration technology. The output current is related to the input voltage by the following equation [15]

T.-Y. Lo and C.-C. Hung, *1V CMOS* G_m-C *Filters: Design and Applications*, Analog Circuits and Signal Processing,

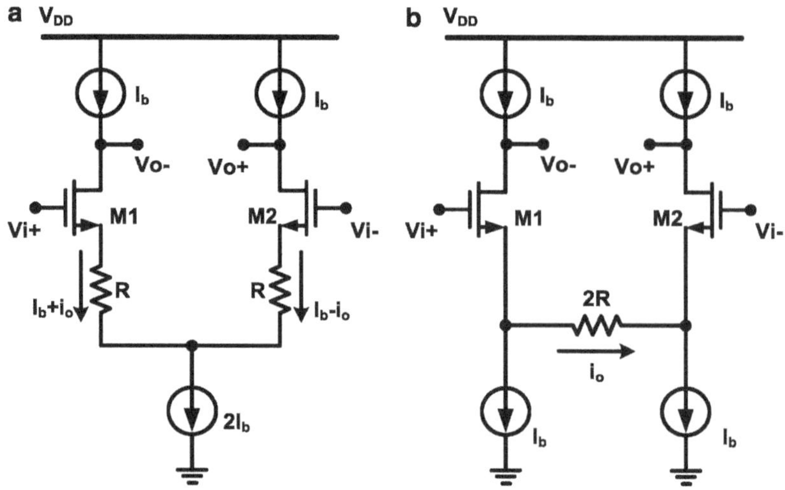

Fig. 2.1 The implementation of the source degeneration transconductor: (**a**) single current source (**b**) dual current source

$$i_O = v_{id}\left(\frac{\sqrt{2K_{(1,2)}I_b}}{1+N}\right)\sqrt{1-\left(\frac{v_{id}}{2\,(1+N)\,V_{DS(1,2)(sat)}}\right)^2} \tag{2.1}$$

$$G_m = \frac{1}{R}\left(\frac{N}{1+N}\right) \tag{2.2}$$

$$HD_3 = \frac{1}{32}\left[\frac{v_{id}}{(1+N)\left(V_{DS(1,2)(sat)}\right)}\right]^2 \tag{2.3}$$

where $v_{id} = Vi{+} - Vi{-}$, $V_{DS(1,2)(sat)} = V_{GS(M1,M2)} - V_{tn}$ and $N = g_{m(1,2)}R$ are the source degeneration factors. From the equation, we can find that the transconductance is reduced by a factor of $1 + N$ and the third harmonic distortion is reduced by the square of the same factor. It is clear that N should be a large value. This condition not only makes the transconductor being tunable, as shown in (2.2), but also increases the linearity performance, as shown in (2.3). The disadvantage of the technique is the higher current and larger aspect ratio when compared with the fully differential pair, and in reality, the required factor of $1 + N$ is expected. The bandwidth of the transconductor is limited due to additional nodes. Fortunately, a zero is obtained in this shunt feedback technology. This zero can be used to compensate the non-dominate high frequency pole. Thus, the topology can still be suitable for high frequency operations, such as a 550 MHz application [16].

Although the circuits in Figs. 2.1a and 2.1b shows the same voltage-to-current relationship, they present different properties. In Fig. 2.1a, the resistor will provide a voltage drop, and then the common-mode voltage range is reduced. In Fig. 2.1b, the noise of the current source will appear at the output, and this source will dominate the noise performance. Mismatch of the current source would also reflect as the input offset.

In many applications, the MOSFET under the linear region is used to replace the passive resistor. The classical NMOS model equation under the linear region is

$$I_D = \mu_n C_{ox}\left(\frac{W}{L}\right)\left[(V_{GS} - V_{tn})\,V_{DS} - a\frac{V_{DS}^2}{2}\right] \tag{2.4}$$

where a is a process dependant parameter. The linear region is hold as the drain-source voltage is lower than the gate-source voltage. If the equation is exact, the perfect linear circuit can be obtained. However, this equation neglects higher order terms under the short channel process, and some of nonlinear terms would occur and then reduce the circuit performance.

If a small drain-source voltage is used, we can simply neglect the second order terms in (2.4). The drain current is linear with respect to the applied drain-source voltage. Thus we can obtain a small-signal resistance of

$$r_{DS} = \frac{1}{\mu_n C_{ox}\left(\frac{W}{L}\right)(V_{GS} - V_{tn})}. \tag{2.5}$$

As the linear region resistor is introduced, the resistance would be proportional to the gate bias voltage and we add the transconductance continuous tuning ability as shown in Fig. 2.1b. The disadvantage of the MOS resistor is the sensitivity of the input common-mode voltage. As we change the input common-mode voltage, the resistance is varied owing to the variation at the source node of transistors M1 and M2. For the reported results, the third-order harmonic distortion (HD3) of MOS resistor within source degenerated transistor is limited to be −50 dB.

Figure 2.2 shows an improved source degeneration transconductor. The circuit relaxes the requirement of large aspect ratio of transistors M1 and M2. However, the feedback topology adds low frequency poles, and thus the circuit is not suitable for high frequency operation.

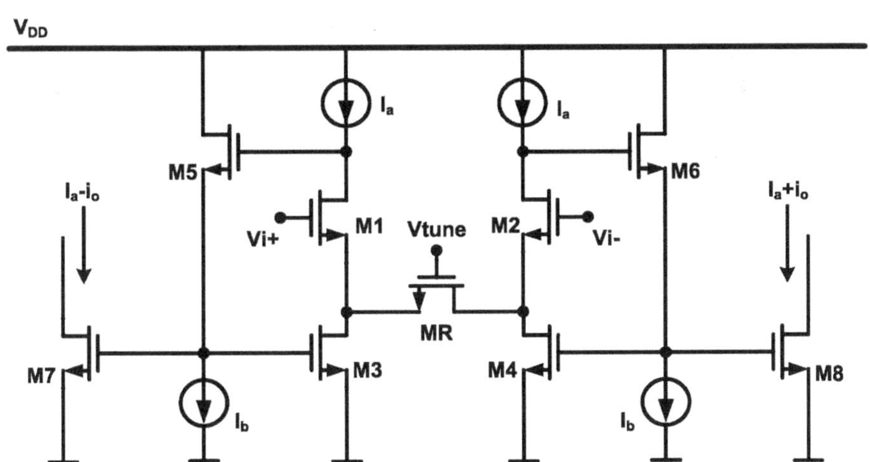

Fig. 2.2 The modified source degeneration transconductor

2.2.2 The Constant Drain-Source Transconductor

In this section, we recall the linear region equation in (2.4) at first. If the drain-source voltage is kept constant, the drain current is linear with respect to the applied gate-source voltage. Figure 2.3 shows the transconductor which uses a constant drain-source technology. By using a regulated control loop, the drain voltage is set to the voltage Vtune through the feedback topology. Then,

$$I_1 = \mu_n C_{ox}\left(\frac{W}{L}\right)\left[(V_{i+} - V_{tn})\,V_{tune} - a\frac{V_{tune}^2}{2}\right] \tag{2.6}$$

$$I_2 = \mu_n C_{ox}\left(\frac{W}{L}\right)\left[(V_{i-} - V_{tn})\,V_{tune} - a\frac{V_{tune}^2}{2}\right] \tag{2.7}$$

We can have,

$$I_1 - I_2 = \mu_n C_{ox}\left(\frac{W}{L}\right)(V_{i+} - V_{i-})\,V_{tune} \tag{2.8}$$

$$G_m = \mu_n C_{ox}\left(\frac{W}{L}\right)V_{tune} \tag{2.9}$$

From the above equation, the transconductance and biasing current are proportional to the Vtune. In practice, the second-order effect would limit the accuracy of the model. In addition, the regulated control loop can relax the requirement of large $gm_{(3,4)}$ to minimize the voltage variation at the drain node of transistors M1 and M2.

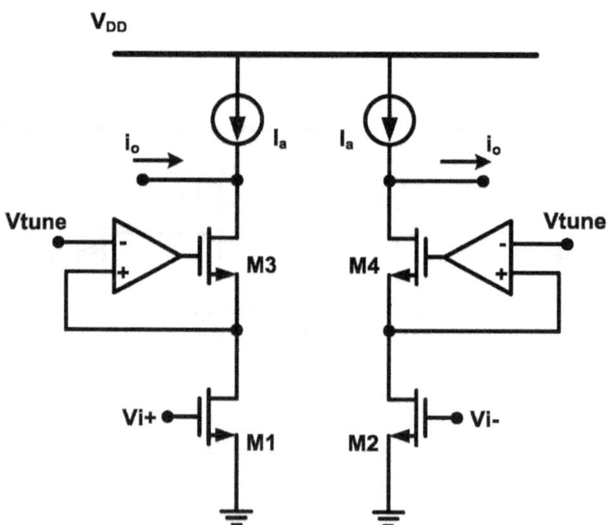

Fig. 2.3 The transconductor by using linear region transistor as inputs

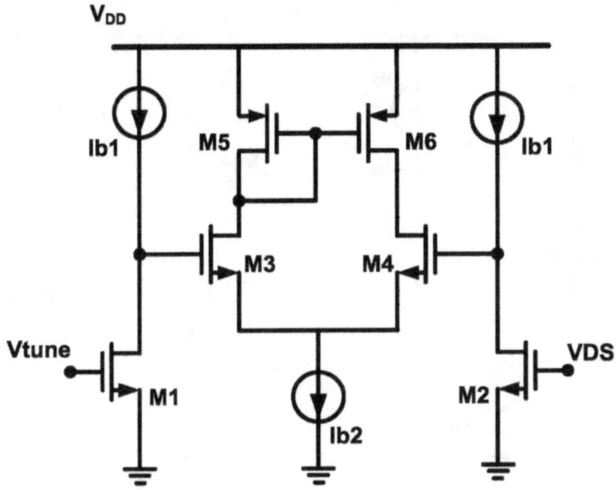

Fig. 2.4 The regulated control loop amplifier

In the circuit, the feedback connection will not appear at the signal path, and thus the transconductor could be used at high frequency as well. In [17], a 200 MHz implementation was presented, and the third order harmonic terms can be expressed as

$$HD3 = \frac{V_i^2 K_{(1,2)}^2}{4\left[(A(s)+1)\,gm_{(3,4)} + K_{(1,2)}(V_{CM} - V_{tn} - V_{tune})\right]^2} \tag{2.10}$$

where A(s) is the gain of the regulated gain control amplifier. In this equation, the high-speed operation would degrade the gain of regulated control amplifier, and thus affect the linearity. From the analysis, a gain of larger than 10 V/V should be required for −50 dB HD3 of the transconductor. The implementation of regulated gain control amplifier is shown in Fig. 2.4. It is composed of differential pair and source followers. The source followers are used to increase the transconductance tuning range and provide more input swing range for the transconductor.

2.2.3 *The Pseudo-Differential Transconductor*

The topology of a general fully-differential transconductor is shown in Fig. 2.5. Figure 2.5a shows the fully-differential transconductor, and Fig. 2.5b shows the pseudo-differential ones. In the fully-differential configuration, the rejection of the common-mode signal is achieved by large output impedance of the tail current source. The transconductor has no internal signal carrying nodes and parasitic poles. Noise from tail current source appears as the common-mode component. The third-order harmonic distortion can be expressed as

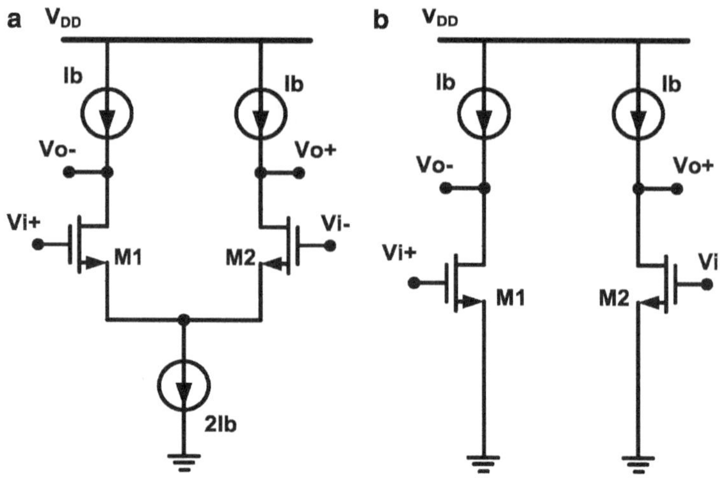

Fig. 2.5 The differential transconductor: (**a**) fully-differential. (**b**) pseudo-differential

$$HD3 = \frac{V_i^2}{32\,(V_{OV} - V_{tn})^2} \tag{2.11}$$

where V_{OV} is the gate overdrive voltage of the input transistors M1 and M2 in Fig. 2.5a. Thus, the linearity can be improved by increasing the gate overdrive voltage of input transistors. For a high linearity transconductor, a value higher than 250 mV is sufficient in the worst case. The way to tune transconductance is to adjust the bias current in the transconductor. However, if a tuning ratio of α is used, we need to increase the bias current with a ratio of α^2 under the saturated square law equation. Thus, we can conclude that only fine tuning is practical.

The pseudo-differential transconductor can be used under low supply voltage because it avoids the voltage drop across the tail current source. We can find that the pseudo-differential structure achieves a larger signal swing. However, the structure presents additional distortion terms produced by the common-mode signal. The even-order terms can appear in a perfectly balanced structure owing to the product of the differential and common-mode signals. Besides, the transconductance of the input common-mode signal is equal to the input differential signal, and it is required to control the input common-mode voltage carefully. Thus, additional common-mode control circuit should be required.

A common-mode feedforward is introduced for input common-mode control in [18]. The common-mode feedforward circuit is shown in Fig. 2.6. The basic idea is to convert the input common-mode through another signal path, and then the cancellation of the input common-mode signal is obtained from the current mirror. In the scheme, transistors M3 and M4 have the same aspect ratio as transistors M1 and M2 to detect the input common-mode signal. Then, the current flowing through transistor M5 is mirrored to cancel the common-mode signal generated by transistors M1 and M2. The common-mode gain is

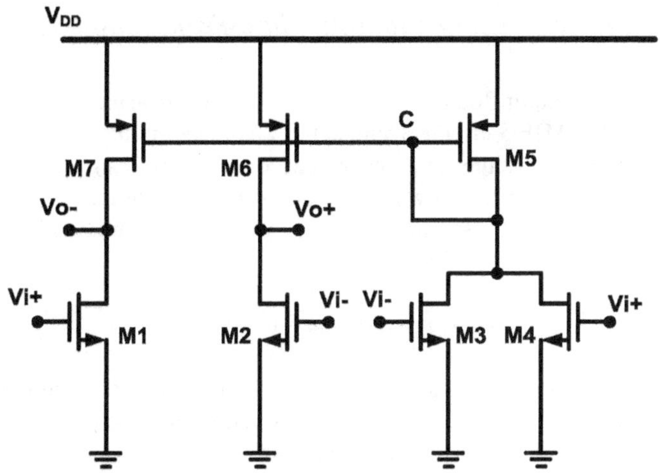

Fig. 2.6 The common-mode feedforward circuit

$$A_{CM} = \frac{g_{m(1,2)}\left(g_{o(1,2)} + g_{o(6,7)} + sC_c\right)}{\left(g_{m(6,7)} + g_{o(1,2)} + g_{o(6,7)} + sC_c\right)\left(g_{o(1,2)} + g_{o(6,7)}\right)} \tag{2.12}$$

where C_c is the parasitic capacitance at node C. At low frequencies, the common-mode gain is approximated to 1. At very high frequencies, the common-mode gain is larger than 1 because the mirroring pole created by node C is grounded. The major drawback of the scheme is larger input capacitance because another differential pair is required.

The pseudo-differential transconductor can be seen as the combination of two parallel transconductors with single ended output. From the ideal square law equation under the saturation region, the output current is perfectly linear with respect to the input voltage. However, the short channel effect would degrade the linearity and the performance is given by

$$HD3 = \frac{`V_i^2}{8V_{OV}\left[1 + \theta\left(V_{OV} - V_{tn}\right)\right]^2\left[2 + \theta\left(V_{OV} - V_{tn}\right)\right]} \tag{2.13}$$

where θ is the mobility reduction coefficient and V_{OV} is the gate overdrive voltage of the input transistors M1 and M2 shown in Fig. 2.5b. Thus, the linearity can also be improved by increasing the gate overdrive voltage of input transistors. For the pseudo-differential transconductor, we can change the transconductance by adjusting the input common-mode voltage. It should be noted that the linearity performance is changed as well.

2.2.4 The Multiple Input Floating-Gate Transconductor

The use of multiple input floating-gate (MIFG) MOS transistors was presented recently [19,20]. The MIFG circuit would act as a voltage divider and thus result in a large linear input swing range. The linear transconductor is designed by the concept of the attenuation. Figure 2.7 shows the block diagram of the technique. The factor α is smaller than 1. We can see that the input voltage is attenuated by the value α and the output current is given by

$$i_o = a_1(\alpha v_i) + a_2(\alpha v_i)^2 + a_3(\alpha v_i)^3 + \cdots \tag{2.14}$$

We can prove that the HD3 gets a value of α^2 smaller than the original one. One of the disadvantages of this technique is the reduction of transconductance and the other is the implementation of perfect attenuator.

The natural attenuation could be obtained from the designed capacitor ratio, and thus the high performance attenuator can be obtained by using the floating gate technique. In the technique, we use multiple input floating gate MOS device. The MIFG MOS transistor is built with a regular MOS transistor where the gate is floating. The basic MIFG NMOS transistor is shown in Fig. 2.8. It can be implemented by double poly process. The floating gate is connected to the input through the capacitor. The equivalent floating gate voltage, V_{FG}, is given by

$$V_{FG} = \frac{C_{GS}V_S + C_{GD}V_D + C_{GB}V_B + \sum_{i=1}^{n} C_i V_i}{C_{GS} + C_{GD} + C_{GB} + \sum_{i=1}^{n} C_i} \tag{2.15}$$

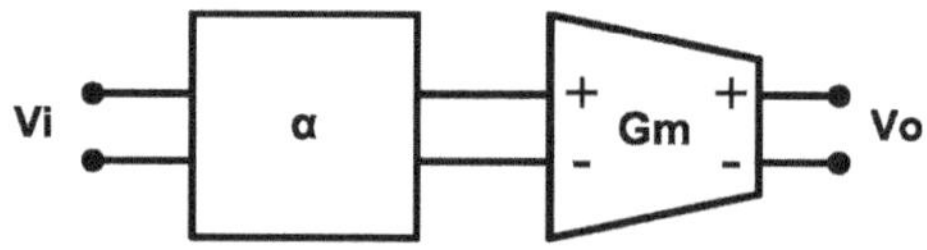

Fig. 2.7 The block diagram of the attenuation technique

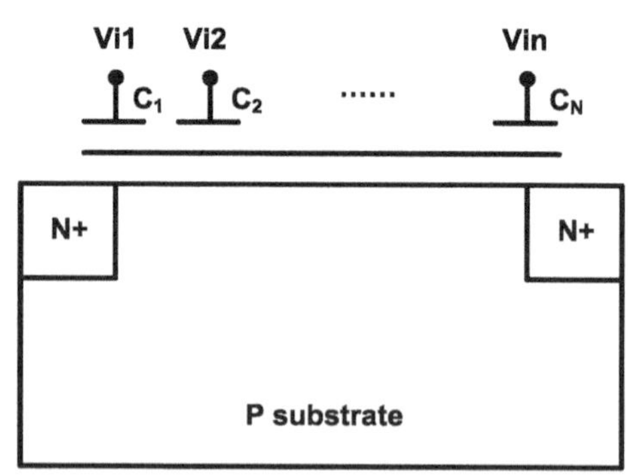

Fig. 2.8 The structure of MIFG NMOS transistor

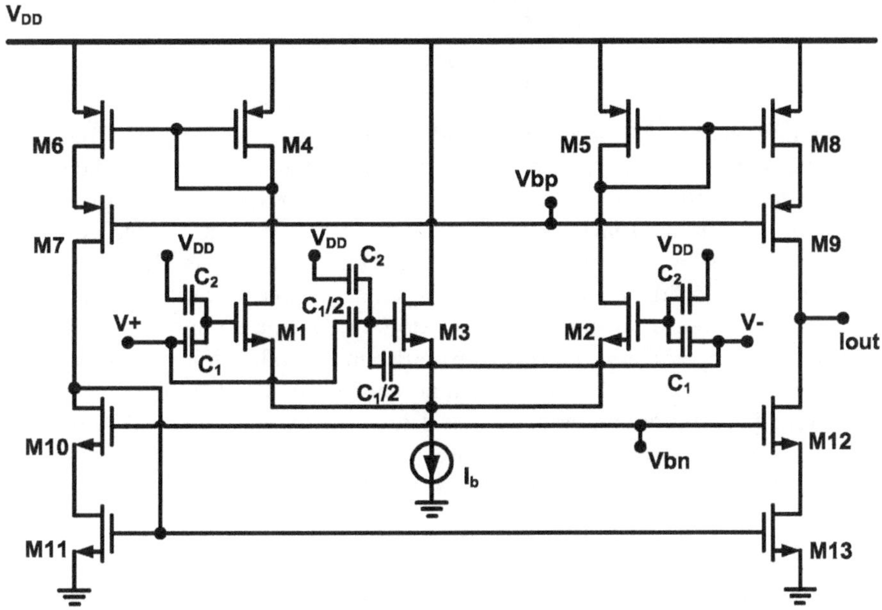

Fig. 2.9 The transconductor implemented by MIFG NMOS transistors

where V_S, V_D, and V_B are the voltages at source, drain, and body nodes of MOS transistor, respectively. The equation shows that the parasitic capacitance will slightly affect the floating gate voltage.

Figure 2.9 shows a linearized transconductor using this technique [19]. The transistors M1 to M3 would be biased at the weak inversion region. Thus, the output current is expressed as

$$I_{out} = \left(\frac{1}{1+A}\right)\left[\frac{w}{2\xi V_T}v_{id} + \left(\frac{1}{6} - \frac{1}{2(1+A)}\right)\left(\frac{w}{2\xi V_T}v_{id}\right)^3 + \left(\frac{1}{120} - \frac{3}{24(1+A)} + \frac{1}{4(1+A)^2}\right)\left(\frac{w}{2\xi V_T}v_{id}\right)^5 + \cdots\right] \tag{2.16}$$

where w is the ratio between the input capacitor and the total capacitance and A = m/2. We note that m is the device parameter ratio of transistor M3 with respect to transistors M1 and M2. Thus, the third-order distortion terms would be cancelled out by choosing A to be 2.

For the transconductance tuning, the MOS capacitor could be a good approach in this technology. The floating gate technique is suitable for low speed applications owing to large input capacitance, and thus the transistors in the weak inversion region are usually used. However, the technique requires extra fabrication processes, and it would not be useful in standard CMOS technology. Moreover, large transconductances would be hardly achieved owing to the voltage attenuation operation.

2.3 The 40 MHZ Double Differential-Pair CMOS Transconductor With −60 DB IM3

A configuration of the linearized transconductor for low-voltage and high frequency applications is proposed. By using double pseudo-differential pairs and the source degeneration structure under nano-scale CMOS technology, the nonlinearity caused by short channel effect from small feature size can be minimized. A robust common-mode control system is designed for input and output common-mode stability, and thus reduces distortion caused by common-mode voltage variation. Tuning ability can be achieved by using MOS transistors in the linear region. The linearity of transconductor is about −60 dB third-order inter-modulation distortion (IM3) for up to 0.9 Vpp at 40 MHz. We note that the definition of IM3 is introduced in Section 3.2.4.4. This transconductor was fabricated by the TSMC 180 nm Deep N-WELL CMOS process. It occupies a small area of $15.1 \times 10^{-3}\,\text{mm}^2$ and the power consumption is 9.5 mW under 1.5-V supply.

2.3.1 Introduction

A variety of linearization techniques have been reported in previous section. However, some of them exploit the ideal square-law behavior of the MOS transistor in the saturation region to obtain high linearity conversion [21, 22]. Unfortunately, this concept is not quite suitable for small feature sizes of MOS transistors due to second-order effects like velocity saturation and mobility reduction. Thus, highly linear transconductors should be designed by taking short channel effects into consideration under nano-scale CMOS technology.

In this section, we present a high linearity and high speed transconductor. It makes use of two input transistor pairs with their source terminals connected to resistor loads and the drain terminals cross-coupled to each other. In the approach, high linearity can be achieved by choosing different values of loading resistors. The model of the short channel effect and the nonlinearity analysis of CMOS transistors are described in Section 2.3.2, and the proposed transconductor implementation is presented in Section 2.3.3. The analyses of non-ideal effects such as mismatch and noise performance of the proposed circuit are discussed in Section 2.3.4. Section 2.3.5 shows the measured performance of fabricated implementation. Finally, the results are summarized in Section 2.3.6.

2.3.2 Nonlinearity Analysis of Saturated MOS Transistors

2.3.2.1 Linearized V-I Characteristics

The relationship between the voltage-to-current conversion can be described as $i_o = f(v_{in})$, where v_{in} and i_o are the input voltage and output current, respectively.

The ideal assumption of the linearized transformation is $f(v_{in}) = k \times v_{in}$, where k is a constant within the applied input voltage range. Unfortunately, the voltage-to-current (V-I) conversion is not possible to be perfectly linear in a real circuit implementation, and the conversion can be investigated by a Taylor series expansion. If the differential structure is applied with well matched implementations, which means the even-order terms can be cancelled out, the current conversion can be expressed as

$$I_O = I_{D1} - I_{D2} = a_1 V_{in} + a_3 V_{in}^3 + a_5 V_{in}^5 + a_7 V_{in}^7 + \cdots \tag{2.17}$$

where the a_i coefficients are determined from the circuit implementation. If the nonlinear factor is suppressed, that is, the parameters $a_{i,i>2}$ are minimized, the V-I conversion would be close to a linear function, as demanded.

2.3.2.2 Saturated MOS Transistor in Nano-Scale CMOS Technology

Linear V-I conversion is usually developed based on the basic square-law behavior of the MOS transistor in the saturation region.

$$I_{D,\ long} = \frac{1}{2} K \left(V_{GS} - V_{thn}\right)^2 \tag{2.18}$$

where $K = \mu_n C_{ox}(W/L)$, W and L are the width and length of the device, respectively, C_{ox} is the oxide capacitance per unit channel area, μ_n is the low-field mobility, and V_{thn} is the NMOS threshold voltage. However, this condition holds only for large length of MOS transistors. As the device size is scaled down towards nano-scale CMOS technology, the short channel effect occurs due to the transverse and longitudinal electric fields. Thus, with the enhancement of speed and area for small device length, the linearity of V-I conversion based on the ideal square-law equation becomes deteriorated. Figure 2.10a shows the circuit of the pseudo-differential input pair. If the length of the MOS transistors is chosen to be the minimum feature size under nano-scale CMOS technology, the output drain current can be modeled by

$$I_{D,\ short} = \frac{K \left(V_{GS} - V_{thn}\right)^2}{2\left[1 + \theta \left(V_{GS} - V_{thn}\right)\right]} \tag{2.19}$$

where θ is the mobility reduction coefficient. From the equation shown above, the mobility reduction coefficient can be modeled by a resistor R_θ connected to the source terminal of an ideal MOS transistor, as shown in Fig. 2.10b. The value of the equivalent resistor equals to θ/K. Moreover, the linearity performance degrades for larger θ. This is confirmed by the results presented in [23], where tunable resistors are introduced in the source terminals of the pseudo-differential pair for the use of transconductance tuning ability with the expense of additional distortion. In this section, in order to resist the nonlinearity occurring from short channel effect, the

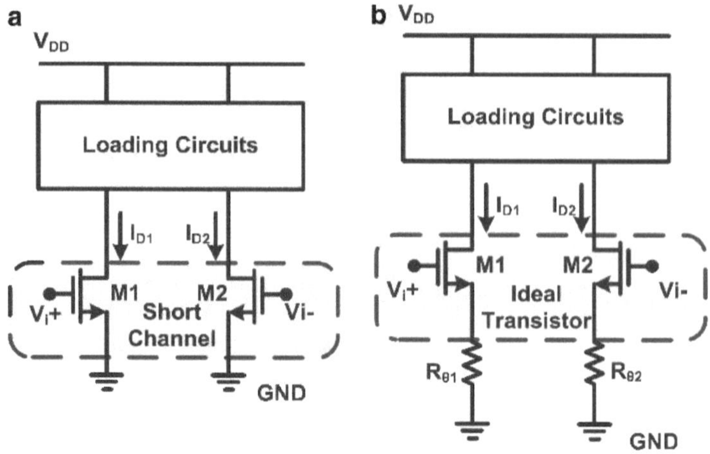

Fig. 2.10 The pseudo-differential circuit by taking short channel effects into consideration: (**a**) short channel transistor (**b**) long channel transistor

Fig. 2.11 Nonlinearity cancellation using double pseudo-differential pairs with degeneration resistors

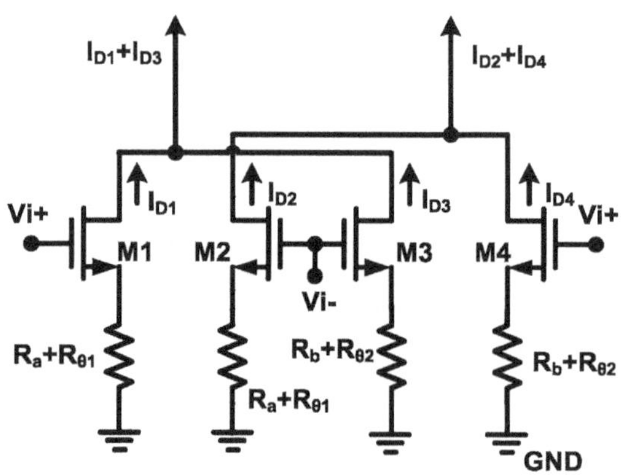

double differential pairs with source degeneration structure are adopted, as shown in Fig. 2.11. In the proposed structure, two different values of resistors, R_a and R_b, are used for each differential pair, and the source degenerated resistors are added up to simplify the expression. Assume that transistors M1 to M4 are operated in the saturation region, and that V_{i+} and V_{i-} are the input differential signals, which would be composed of common-mode and differential-mode voltages

$$
\begin{aligned}
V_{i+} &= V_{cm} + \frac{V_{id}}{2} \\
V_{i-} &= V_{cm} - \frac{V_{id}}{2}
\end{aligned}
\tag{2.20}
$$

where V_{cm} is the input common-mode voltage and V_{id} is the input differential-mode voltage. Then, the output current of each transistor could be given by

$$\begin{aligned} I_{D1} &= \frac{K_1 (V_{i+} - V_{thn})^2}{2\left[1 + K_1 \left(R_a + R_{\theta_1}\right)(V_{i+} - V_{thn})\right]} \\ I_{D2} &= \frac{K_2 (V_{i-} - V_{thn})^2}{2\left[1 + K_2 \left(R_a + R_{\theta_2}\right)(V_{i-} - V_{thn})\right]} \\ I_{D3} &= \frac{K_3 (V_{i-} - V_{thn})^2}{2\left[1 + K_3 \left(R_b + R_{\theta_3}\right)(V_{i-} - V_{thn})\right]} \\ I_{D4} &= \frac{K_4 (V_{i+} - V_{thn})^2}{2\left[1 + K_4 \left(R_b + R_{\theta_4}\right)(V_{i+} - V_{thn})\right]} \end{aligned} \tag{2.21}$$

where $K_1 = K_2$, $K_3 = K_4$, $R_{\theta 1} = R_{\theta 2}$, and $R_{\theta 3} = R_{\theta 4}$. Thus, under ideal matching, the differential output current would be the function of the input signals

$$\begin{aligned} I_o &= (I_{D1} + I_{D3}) - (I_{D2} + I_{D4}) = f(V_{id}) \\ &= \left(a_{1,(1,2)} - a_{1,(3,4)}\right) V_{id} + \left(a_{3,(1,2)} - a_{3,(3,4)}\right) V_{id}^3 + \cdots \end{aligned} \tag{2.22}$$

where $a_{j,i}$ is the jth-order harmonic component provided by the ith transistor of the proposed structure, $a_{j,1} = a_{j,2} = a_{j,(1,2)}$, and $a_{j,3} = a_{j,4} = a_{j,(3,4)}$. Although the resistors connected to the source of a single pseudo-differential pair degrade the linearity performance, the third-order harmonic component could be cancelled out by proper sizing of the double pseudo-differential pairs through a Taylor series expansion of (2.22)

$$a_{3,(1,2)} - a_{3,(3,4)} = 0 \tag{2.23}$$

This expression can be obtained by taking

$$\frac{\left(W_{(1,2)}/L_{(1,2)}\right)^2 \left(R_a + R_{\theta(1,2)}\right)}{\left[2 + \left(R_a + R_{\theta(1,2)}\right) g_{m(1,2)}\right]^4} = \frac{\left(W_{(3,4)}/L_{(3,4)}\right)^2 \left(R_b + R_{\theta(3,4)}\right)}{\left[2 + \left(R_b + R_{\theta(3,4)}\right) g_{m(3,4)}\right]^4} \tag{2.24}$$

where W_i, L_i, and g_{mi} are the width, length, and transconductance of the ith transistor, respectively, and $R_{\theta i}$ is the ith short channel equivalent resistance. To minimize of the third-order harmonic component, the transconductance of the proposed structure is given by

$$\begin{aligned} G_{m,total} = &\frac{g_{m(1,2)} \left[2 + \left(R_a + R_{\theta(1,2)}\right) g_{m(1,2)}\right]}{2\left[1 + \left(R_a + R_{\theta(1,2)}\right) g_{m(1,2)}\right]^2} \\ &- \frac{g_{m(3,4)} \left[2 + \left(R_b + R_{\theta(3,4)}\right) g_{m(3,4)}\right]}{2\left[1 + \left(R_b + R_{\theta(3,4)}\right) g_{m(3,4)}\right]^2} \end{aligned} \tag{2.25}$$

The transconductance decreases owing to the use of double differential pairs and source degeneration resistors. This implies higher linearity with the tradeoff of higher power consumption.

2.3.2.3 Design Methodology

In order to obtain high linearity performance for the double differential pair structure under optimal transconductance efficiency, a simple approach is used by giving the ratio of parameters to represent the circuit operation. Thus, by taking

$$\frac{W_{(1,2)}/L_{(1,2)}}{W_{(3,4)}/L_{(3,4)}} = P \quad \frac{R_a + R_{\theta(1,2)}}{R_b + R_{\theta(3,4)}} = Q \quad V_{cm} = \frac{V_{DD}}{2} \tag{2.26}$$

we can find that the ratios of (2.26) would be used to define the transconductance efficiency in comparison with the single differential pair circuit and the third-order harmonic component of the proposed circuit. In addition, the ratio values should be designed within the practical implementation boundary. Bandwidth, noise performance, and matching are also taken into consideration.

The optimization procedure starts from the reduced transconductance value. We define that less then 30% of the transconductance should be reduced with respect to that of a single differential pair circuit with the same size and current consumption. From Fig. 2.12, we can find that if the value of Q is set to 3 under large P, we can obtain less than 30% reduction in the transconductance. Moreover, if a value of 4 is used for Q, less than 25% reduction in the transconductance would be obtained.

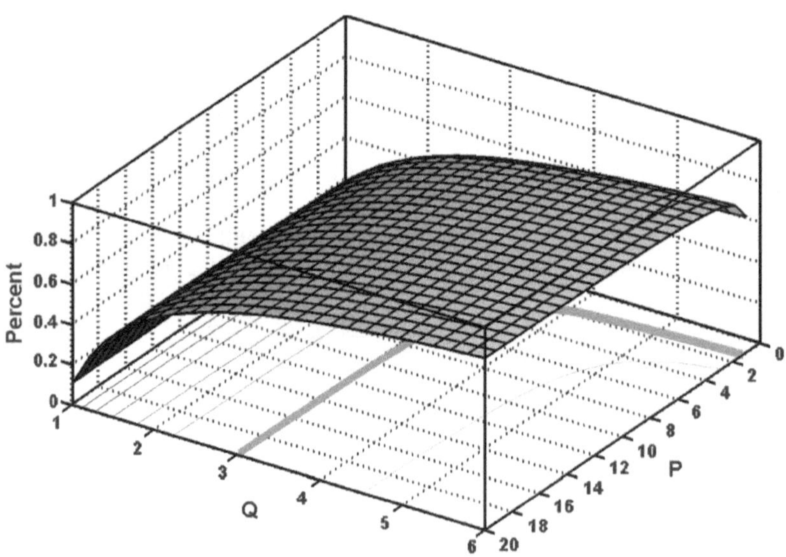

Fig. 2.12 Optimal parameter evaluation for the reduced transconductance

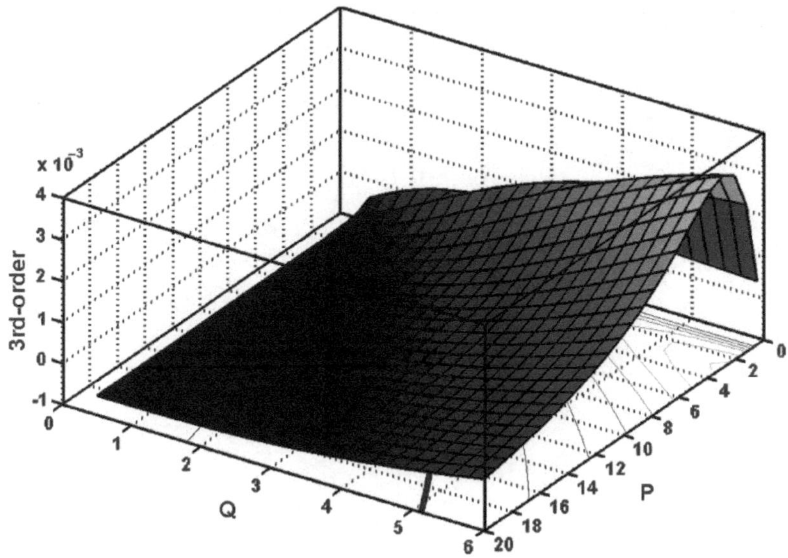

Fig. 2.13 Optimal parameter evaluation for the third-order harmonic component

Figure 2.13 shows the third-order distortion component of the proposed design. In order to obtain the minimized distortion components, the ratio P is chosen as 9 while Q is set to be 3. If Q is set to 4 for less transconductance reduction, P should be set to 16, but such a large ratio would degrade the bandwidth performance owing to the large parasitic capacitance of input transistors. After the optimization procedure, the optimal ratios of the proposed circuit is given by

$$P = 9 \quad Q = 3 \tag{2.27}$$

The optimization procedure concludes that the third-order distortion component is ideally cancelled out with the expected transconductance value, as shown in Fig. 2.14. The linearity performance is actually robust to process variation owing to the flat distribution in Fig. 2.13. Thus, the small reduction in the transconductance value makes high linearity and high speed possible under about 30% of extra power consumption.

Transconductance tuning is another important issue in the transconductor design. The main idea of the transconductance tuning is to compensate for the variation caused by the fabrication process and temperature. Figure 2.14 shows the contour plot of the third-order harmonic component under transconductance tuning, resulting from Fig. 2.13. We can find that if Q is changed from 1 to 4 when P is set to 9, it implies more than 300% of the transconductance tuning range, as shown in Fig. 2.12, and the third-order harmonic component value of less than 0.001 can be guaranteed, as illustrated in Fig. 2.14.

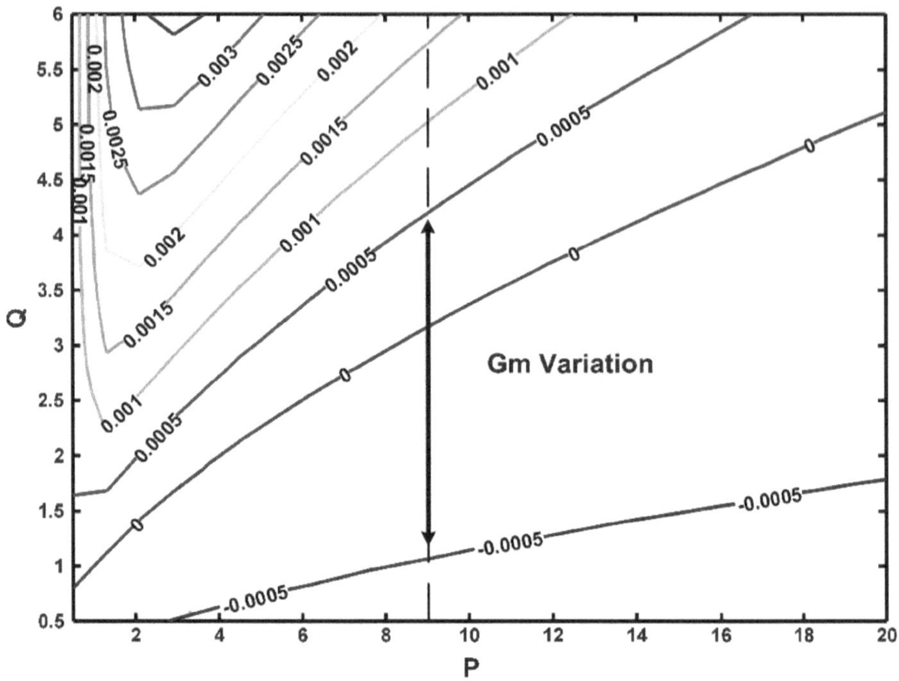

Fig. 2.14 Contour plot for the third-order harmonic component under transconductance tuning

2.3.3 *Proposed Transconductor Circuit*

2.3.3.1 Implementation of Linearization Technique

Figure 2.15 shows the proposed transconductor design. Two differential pairs M1 to M2 and M3 to M4 are used in order to cancel the nonlinearity component, as described in the previous section. For continuous transconductance tuning strategy, transistors M5 to M8 operating in the linear region are used to replace the resistors. The equivalent resistance is given by

$$R_{eq}^{-1} = K(V_G - V_{thn}) \tag{2.28}$$

where V_G is the gate voltage of the transistor. Therefore, we can obtain the required equivalent resistance by applying the voltages V_a and V_b.

$$R_a^{-1} = K_{(5,6)}(V_a - V_{thn}) \tag{2.29}$$
$$R_b^{-1} = K_{(7,8)}(V_b - V_{thn}) \tag{2.30}$$

The linearity can be maintained by proper sizing of the degenerated transistors and the control voltage. In addition, the tuning ability of the proposed circuit can be achieved by adjusting the control voltages V_a and V_b. Figure 2.16 shows the

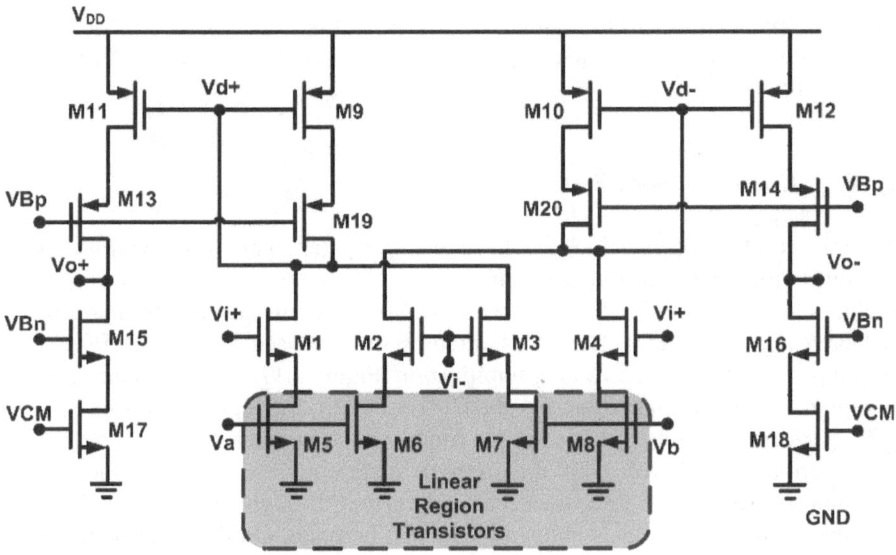

Fig. 2.15 Proposed transconductor circuit

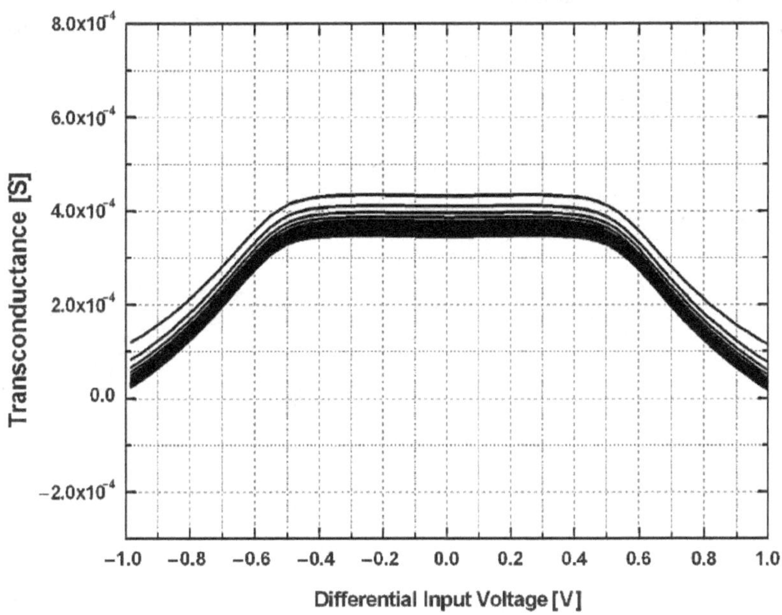

Fig. 2.16 Simulated transconductance tuning range

simulated large-signal transconductance of the differential transconductor operating on a supply voltage of 1.5-V. The proposed circuit can be tuned from 360 μS to 470 μS. It can be noticed that the transconductance tuning range is limited by the linear-region operation of transistors M5 to M8. Besides, the speed of the proposed transconductor is mainly limited by the parasitic capacitors caused by the current mirror circuits.

2.3.3.2 Common-Mode Stability

The transconductor shown in Fig. 2.15 requires a proper common-mode control system due to the pseudo-differential structure [24]. The common-mode control system includes the common-mode feedforward (CMFF) circuit and the common-mode feedback (CMFB) circuit. The CMFF circuit should be used with the CMFB circuit for output common-mode voltage stabilization. Figure 2.17 shows the circuit of the common-mode control system. For the CMFB circuit, the input transistors MF1 to MF4 perform the tasks of the common-mode detection and reference comparison. If the common-mode voltage of the transconductor output signal equals the desired common-mode voltage Vref, then the total current through MF7 will be constant and the common-mode bias voltage VCM will be fixed. On the other hand, if the common-mode voltage of the transconductor output signal is not the same as Vref, a current will be mirrored by MF9 to change VCM adaptively. Thus, the feedback mechanism adjusts the output common-mode voltage to the desired value.

Furthermore, the input common-mode control circuitry is formed by transistors MF14 to MF18 which constitute the CMFF circuit. The combination of transistors

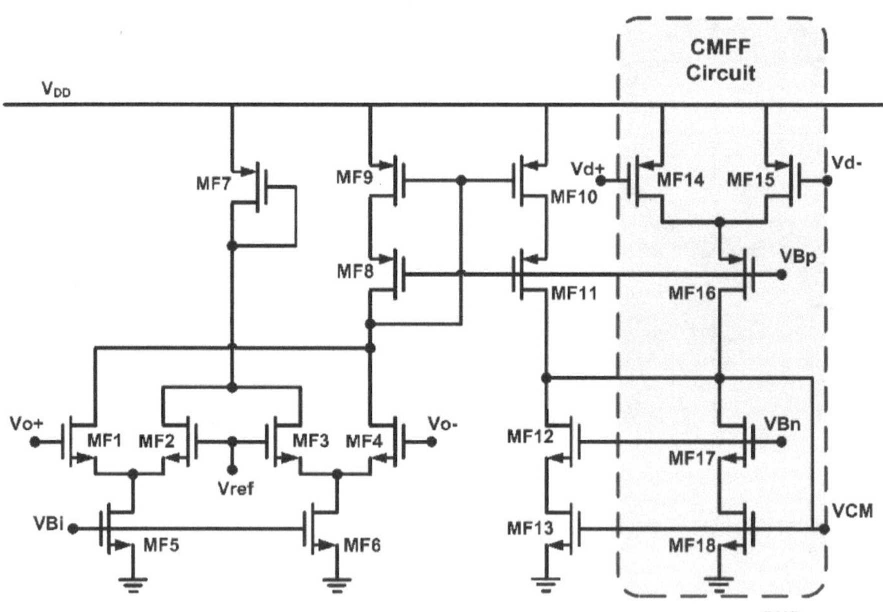

Fig. 2.17 The common-mode control system

MF14 and MF15 generates a scaled copy of input common-mode currents, which is subtracted at the transconductor output stage through the use of current mirror MF17 to MF18. Thus, the input common-mode signal could be suppressed out and only the differential-mode signal appears at the output stage. As the mechanism shown above, it is demonstrated that the common-mode control circuit can be implemented to achieve excellent stability over the tuning range. Moreover, linearity could be maintained by the robust and stable common-mode control system.

The common-mode rejection (CMR) depends on matching. We can define a matching factor of $(1 + \Delta)$ between the CMFF path and the signal path, where Δ is the mismatch ratio. We can emulate the CMFB circuit as a small resistor of value $1/g_{CMFB}$, and then the common-mode gain of $A_{CM} = G_{m,total}(g_o - \Delta \times g_{m(17,18)})/(g_{m(17,18)} \times g_{CMFB})$ at low frequency can be obtained, where $G_{m,total}$ is obtained from (2.25), $g_{m(17,18)}$ is the transconductance of transistors M17 and M18, and g_o is the output conductance of the transconductor. This is the result of combined CMFB and CMFF systems. Since g_{CMFB} is large, A_{CM} is much less than unity even mismatch problems occur so that high CMR can be obtained.

2.3.4 *Non-Ideality Analysis of the Implementation*

2.3.4.1 Mismatch

Owing to the non-ideal matching phenomena of MOS transistors, the nonlinearity cancellation is not perfect and second-order harmonic distortion components would still appear at the differential output nodes. For the double differential pair structure, it is assumed that there are mismatches of $K_{(1.2)} \pm \varepsilon_a K_{(1,2)}$ for transistors M1 and M2 and $K_{(3,4)} \pm \varepsilon_b K_{(3,4)}$ for transistors M3 and M4. Repeating the analysis of (2.22), the second-order distortion component resulting from mismatch is given by

$$a_2 \approx \frac{\varepsilon_a K_{(1,2)}}{\left[1 + \left(R_a + R_{\theta(1,2)}\right) g_{m(1,2)}\right]^3} + \frac{\varepsilon_b K_{(3,4)}}{\left[1 + \left(R_b + R_{\theta(3,4)}\right) g_{m(3,4)}\right]^3} \tag{2.31}$$

Therefore, the distortion components caused by transistor mismatch could be minimized by applying large degenerated resistors and gate overdrive voltage. Besides, the current mirrors M9 to M12 would also contribute second-order distortion components under the proposed degenerated structure, and thus large device sizes and small aspect ratios could be designed. From the simulation with 2% transistor mismatch, the highest even-order components remain lower than odd-order components by at least 5 dB. In addition, careful layout was taken while the device match is required. The error output current contributed by transistor mismatch can be divided by the overall transconductance to model an equivalent offset voltage, and it could be removed by applying an offset voltage of input differential signals.

2.3.4.2 Thermal Noise

For the high-speed circuit, the most significant noise source of a single transistor is the thermal noise rather than the flicker noise. The channel noise can be modeled by a current source connected between the drain and source with a spectral density

$$\overline{I_n^2} = 4kT\delta g_{ms} \tag{2.32}$$

where k is the Boltzmann constant, T is the absolute temperature, g_{ms} is the source conductance, and the device noise parameter δ depends on the bias condition [25]. Using the thermal noise model, the total output-referred noise spectral density of the double differential pairs with degeneration structure can be derived as

$$\begin{aligned}\overline{I_{n,out}^2} \approx 8kT \Bigg[&\delta_s g_{m(9,10)} + \delta_s g_{m(1,2)} \left(\frac{1}{1 + g_{m(1,2)} R_a}\right)^2 \\ &+ \delta_l R_a \left(\frac{g_{m(1,2)}}{1 + g_{m(1,2)} R_a}\right)^2 + \delta_s g_{m(3,4)} \left(\frac{1}{1 + g_{m(3,4)} R_b}\right)^2 \\ &+ \delta_l R_b \left(\frac{g_{m(3,4)}}{1 + g_{m(3,4)} R_b}\right)^2 \Bigg] \left(\frac{g_{m(11,12)}}{g_{m(9,10)}}\right)^2 + 8kT\delta_s g_{m(17,18)}\end{aligned} \tag{2.33}$$

where δ_s and δ_l are the noise parameters at saturation and linear regions, respectively. The input-referred noise spectral density can be calculated by dividing the output-referred noise spectral density by the overall transconductance. From the noise analysis, large aspect ratios of input transistors and small aspect ratios of load transistors should be designed. The input-referred noise of the proposed circuit is higher than the single differential pair circuit owing to the fact that the noise contribution is the combination of two input differential pairs. Moreover, the degenerated MOS resistors contribute additional noise sources to the proposed circuit.

2.3.5 Experimental Results

The proposed transconductor has been fabricated using the TSMC 180 nm Deep N-WELL CMOS process. Its operation and linear V-I characteristics have been studied. A micrograph of the linear transconductor is depicted in Fig. 2.18 and the occupied area is $15.1 \times 10^{-3}\,\text{mm}^2$. A supply voltage of 1.5-V is employed in the measurements and the nominal static power consumption of the transconductor is 9.5 mW. The required supply voltage for the circuit is $V_{GS} + 2\,V_{DS(\text{saturation region})}$, and 1.5-V is sufficient for this circuit to operate under 180 nm CMOS process.

For the measurement setup, the output signal of the signal generator was passed through a low-pass filter for the spectral purity of the input signal. The transformers

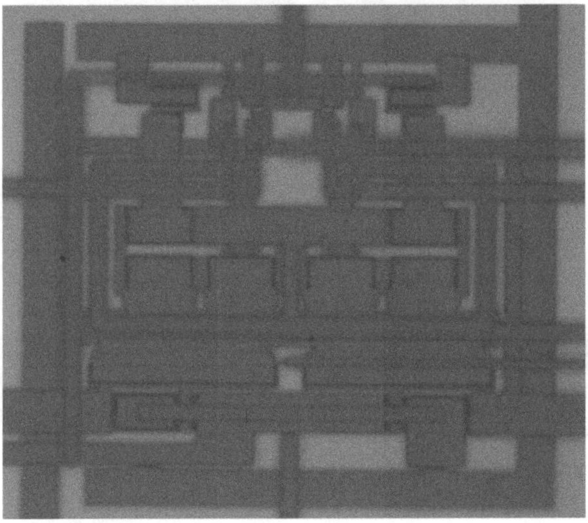

Fig. 2.18 Die microphotograph

were used before and after the input and output terminals for single-to-differential and differential-to-single conversion for the differential circuit. The output signal was measured with a spectrum analyzer. The third-order inter-modulation distortion measured with two sinusoidal tones of 0.9 V_{pp} amplitude is shown in Fig. 2.19. The IM3 is shown to be about −60 dB at a speed of 40 MHz. Figure 2.20 shows the nonlinearity behavior with respect to the frequency under the same input swing range. At low frequencies, the IM3 of −75 dB could be obtained. Moreover, IM3 less than −55 dB could be achieved for frequency up to 60 MHz. The increment in IM3 is due to the different high frequency behaviors of the two input differential pairs. The measured input referred noise spectral density at 40 MHz is 23 nV/$\sqrt{\text{Hz}}$. Table 2.1 summarizes this work with those reported recently. In order to compare with different transconductor implementations, the defined figure of merit (FOM), which takes the transconductance value, linearity performance, speed of the implemented circuit, input swing range, and power consumption into account, is expressed as

$$FoM = 10\log\left(\frac{G_m \times V_{id} \times IM3_{linear} \times f_o}{power}\right) \tag{2.34}$$

Therefore, our high speed linear transconductor compares favorably with the literature.

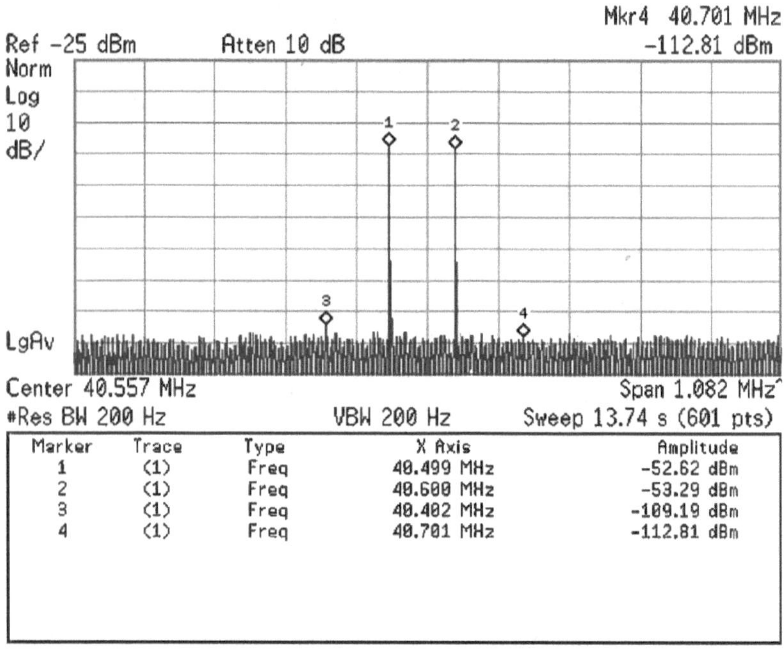

Fig. 2.19 Measured two tone inter-modulation distortion

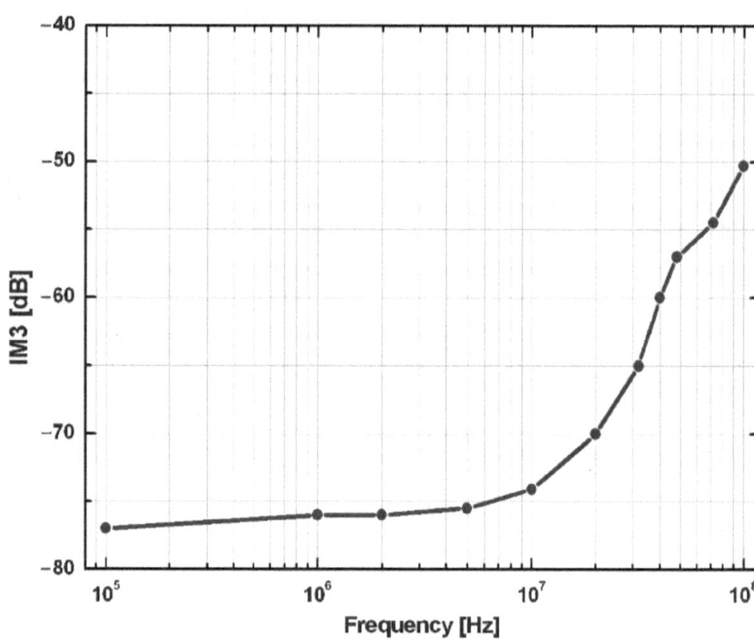

Fig. 2.20 Measured two tone inter-modulation distortion with respect to input signal frequency

Table 2.1 Comparison with previously reported works

Reference	2005 TCAS-I [19]	2005 TCAS-I [22]	2003 JSSC [26]	2006 TCAS-II [27]	2004 TCAS-II [28]	2005 JSSC [29]	This work
Technology	0.8 μm CMOS	0.8 μm CMOS	0.5 μm CMOS	0.18 μm CMOS *Simulation	0.35 μm CMOS	0.5 μm CMOS	0.18 μm CMOS
Trancondu ctance Value	0.08μS	266μS	1065μS	20μS	100μS	100μS	470μS
Linearity	-40dB THD at 100Hz	-43dB HD3 at 1kHz	-43dB HD3 at 30MHz	-65dB HD3 at 1MHz	-65dB IM3 at 20MHz	-66.5dB THD at 100kHz	-60dB IM3 at 40MHz
Input swing range	1.1 V_{PP}	0.4 V_{PP}	0.9V_{PP}	0.6V_{PP}	1.3V_{PP}	2V_{PP}	0.9V_{PP}
Supply Voltage	1.5V	2V	3.3V	1.8V	3.3V	2.6V	1.5V
Power consumption	1μW	150μW	10.7mW	145μW	10.5mW	1.7mW	9.5mW
Figure of merit (FOM)	29	50	84	82	86	74	93
Input-referred noise spectral density	-	-	9.8nV/√Hz	-	75nV/√Hz	-	23nV/√Hz

2.3.6 Summary

An approach to enhance the transconductor linearity under nano-scale technology has been proposed, and the experimental result proves the same linear characteristics by the fabricated chip. By taking the short channel effect into consideration under small feature sizes, this approach is based on the nonlinearity cancellation scheme with two pseudo differential pairs of the source degeneration structure, and the circuit works well at high frequencies. The MOS transistors working in the linear region were used to replace the poly resistors. It not only saves the chip area but also adds the tuning ability. A common-mode control circuitry, including the CMFF and CMFB circuits, is used for the input and output common-mode stability. The measurement results show about −60 dB IM3 with 40 MHz 0.9 V_{pp} input signals under a 1.5-V supply.

2.3.7 Analysis for Circuit at 1-V Supply

To maintain correct operation region for transistors in this circuit, the input signal should be larger than NMOS threshold voltage and smaller than bias voltage V_a and V_b. Since the bias voltage is usually limited by supply voltage, the input common-mode voltage will be degraded as well. When the supply voltage is reduced to 1-V, the upper boundary of the signal range is decreased, and the smaller bias voltage should be used. This effect leads to larger triode region MOS resistors and smaller transistor transconductance. Therefore, the overall transconductance in this circuit would be decreased. However, when the process is scaled down to 130 nm CMOS, smaller threshold voltage could be obtained, and thus the lower boundary of the signal range is extended. Figure 2.21 shows the simulated transconductance tuning range for 1-V supply by using a 130 nm CMOS process spice model. This figure shows that the linearity and the tuning range of the simulated transconductance can be well maintained in this circuit. Although smaller transconductance is obtained, the power is also scaled down at the same time. In addition, we can find that the gain of this transconductor would be increased for smaller bias voltage and thus CMRR performance is maintained. Simulation result shows the value at low frequency is 39 dB. Therefore, we can conclude the proposed circuit also operates well at 1-V supply.

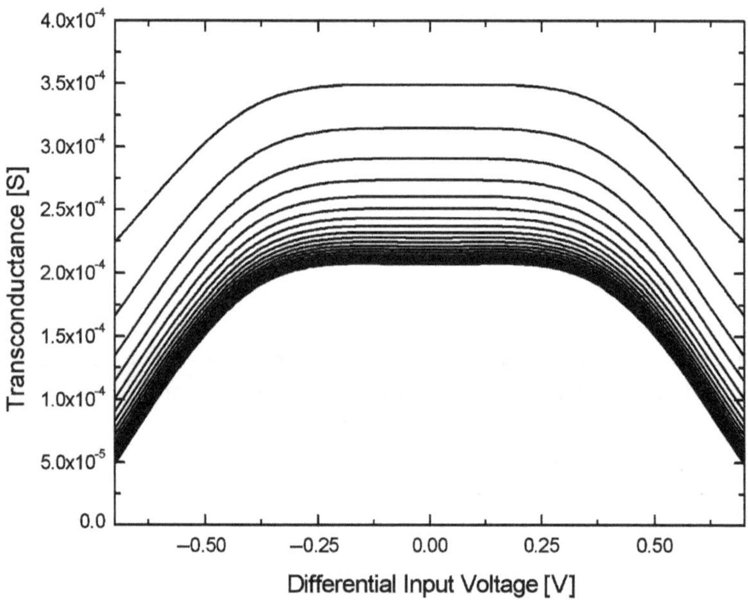

Fig. 2.21 Simulated transconductance tuning range at 1-V supply

2.4 A 50 MHZ Pseudo-Differential Transconductor With Mobility Reduction Compensation

This section presents a high linearity transconductor based on pseudo-differential structures. The linearity is improved by mobility compensation techniques as the device size is scaled down to achieve high-speed operation. Transconductance tuning could be achieved by a MOS operating in the linear region. The transconductor fabricated in the 180 nm CMOS process occupies a small area of $4.5 \times 10^{-3}\,\mathrm{mm}^2$. The measured third-order inter-modulation distortion with a 400 $\mathrm{mV_{pp}}$ differential input under 1-V supply remains below −52 dB for frequency up to 50 MHz. The static power consumption is 2.5 mW. The experimental results demonstrate good agreement with theoretical analyses.

2.4.1 Introduction

The circuit that converts the applied input voltage into current at output nodes is generally referred to as a V-I converter or transconductor. The transconductor is a basic building block in analog VLSI applications, including continuous-time filters and four-quadrant multipliers. The precision of the data signal processing is limited by the performance of the linearity and noise. There are numerous previous works to improve the transconductor linearity. On one hand, with the reduction of supply voltage, the operating range is decreased. On the other hand, linearity based on the MOS transconductance or saturated square-law behavior becomes worse in the sub-micron process owing to the appearance of short channel effects. The combination of linearization techniques has been used recently [30–32], but the potential of getting higher power consumption makes the combined linearity mechanism less practical. Therefore, maintaining high linearity while achieving high speed becomes very challenging under the conditions of low supply voltage and limited power consumption.

This section presents a highly linear transconductor at 1-V supply. The design issues and circuit details are discussed in Section 2.4.2. Compensation of the nonlinearity is also analyzed in this section. The experimental results are presented in Section 2.4.3. Finally, Section 2.4.4 summarizes the implementation.

2.4.2 The Proposed Transconductor Cell

Figure 2.22 shows the block diagram of the proposed approach. It is known that under the differential architecture, the third-order distortion of the transconductor will be the dominant term in its nonlinearity performance. Two voltage-to-current converters with the same first-order sign and opposite third-order sign are provided

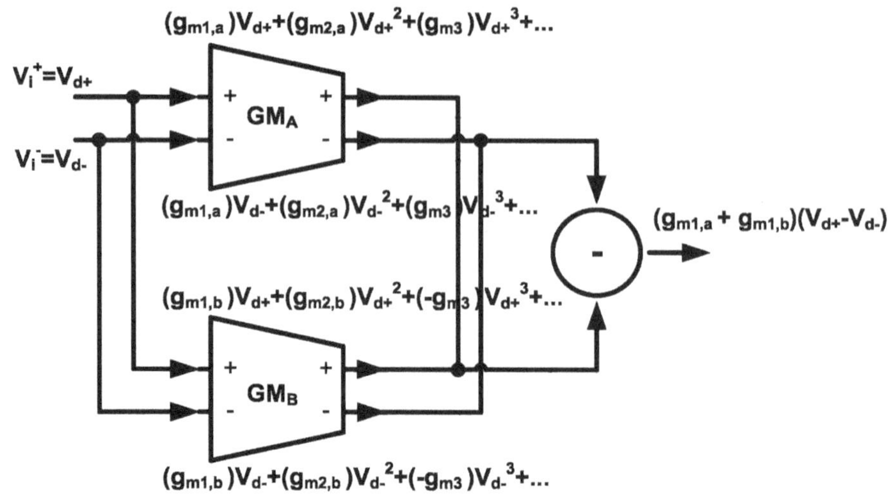

Fig. 2.22 Nonlinearity cancellation mechanism

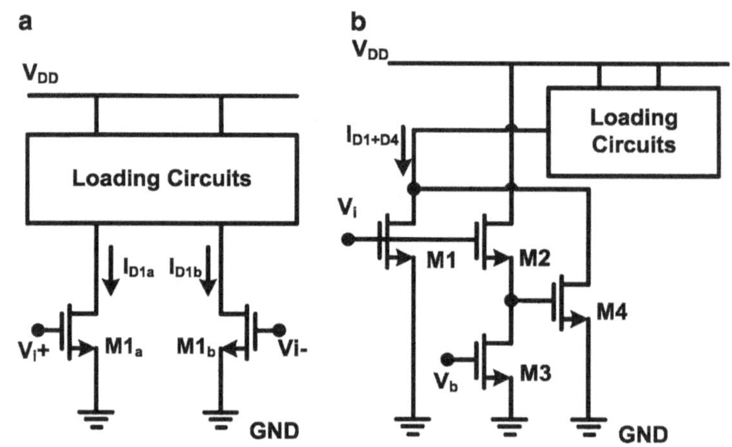

Fig. 2.23 Mobility compensation technique: (**a**) basic pseudo-differential CMOS pair (**b**) mobility compensation in the pseudo-differential structure

in the transconductor. Therefore, the nonlinearity term of the transconductor can be reduced with the addition of two transconductance.

2.4.2.1 Mobility Compensation

The circuit implementation of the block diagram is based on the fact that transistors working in the saturation and weak inversion regions have opposite signs for the third-order harmonic distortion. Figure 2.23a shows the basic pseudo-differential CMOS pair. When transistors $M1_a$ and $M1_b$ operate in the saturation region, the effective carrier mobility would be the function of longitudinal and transversal electric fields due to short channel effects. The drain current can be approximated as

$$I_{D,sat} = \frac{\mu_n C_{ox}\left(\frac{W_S}{L_S}\right)(V_{GS} - V_{thn})^2}{2\left[1 + \theta\left(V_{GS} - V_{thn}\right)\right]}(1 + \lambda V_{DS}) \tag{2.35}$$

where W_s and L_s are the width and length of the device, respectively, C_{ox} is the oxide capacitance per unit channel area, μ_n is the low-field mobility, θ is the mobility reduction coefficient, V_{thn} is the NMOS threshold voltage (a typical value of 0.43 V in the adopted process), and λ is the output impedance constant. By taking nonlinear effects into account for the saturated device, the differential output current can be expanded by Taylor series as

$$I_O = I_{D1} - I_{D2} = a_1 V_{in} + a_3 V_{in}^3 + a_5 V_{in}^5 + a_7 V_{in}^7 + \cdots \tag{2.36}$$

where V_{in} equals to ($V_i+ - V_i-$) and the even-order harmonic distortion terms are cancelled out due to the topology of the differential structure. The third-order harmonic distortion term of the saturated transistors can be calculated as

$$a_{3,sat} = -\frac{\mu_n C_{ox}\left(\frac{W_s}{L_s}\right)\theta}{32\left[1 + \theta\left(V_{cm} - V_{thn}\right)\right]^4} \tag{2.37}$$

where V_{cm} is the input common-mode voltage. The above equation is inversely proportional to the input common-mode voltage, so the third-order harmonic distortion term increases rapidly for low supply voltage. On the other hand, the drain current of the weak inversion transistors can be expressed as

$$I_{D,sub} = I_0\left(\frac{W_w}{L_w}\right)e^{\frac{V_{GS}}{\xi V_T}}\left(1 - e^{\frac{-V_{DS}}{V_T}}\right) \tag{2.38}$$

where W_w and L_w are the width and length of the device, respectively, I_o is the process-dependent parameter, ξ is the weak inversion slope factor, and V_T is the thermal voltage. When transistors $M1_a$ and $M1_b$ work in the weak inversion region, the coefficient of the third-order harmonic distortion term of the differential output current is given by

$$a_{3,sub} = \frac{I_0}{24\left(\xi V_T\right)^3}\left(\frac{W_w}{L_w}\right)e^{\frac{V_{cm}}{\xi V_T}} \tag{2.39}$$

The third-order harmonic distortion term decreases with smaller device width and smaller input common-mode voltage. As the current in the saturated transistor is combined with that of the weak inversion transistor, the nonlinearity term can be reduced by choosing proper aspect ratios such that

$$a_{3,sat} + a_{3,sub} = 0 \tag{2.40}$$

Figure 2.23b shows the implementation of the concept. Transistors M1, M2, and M3 are working in the saturation region. The source follower with M2 and M3 is used to force M4 to enter the weak inversion region. Thus, the third-order nonlinear term can be reduced by

$$a_{3,sat} + \frac{1}{\alpha} a_{3,sub} = 0 \tag{2.41}$$

where α is inversely proportional to the gain of the source follower. Therefore, the third-order harmonic distortion term will be cancelled out by adding drain current of M1 and M4 together. Since transistor M4 needs to work in the weak inversion region in order to reduce the harmonic distortion, the source follower plays an important role in the approach. For large bias current, the gate-source voltage of the source follower will become larger.

In this case, the minimum input swing range and the aspect ratio of transistor M4 should be increased in order for M4 to stay in the weak inversion region and for effective nonlinearity compensation. If a smaller source follower gain is maintained, the source follower would work as a voltage attenuator and the input swing range can be increased. However, under this condition, the compensation ability of the third-order harmonic distortion provided by the weak inversion transistor M4 will become weaker due to a higher attenuation ratio. Therefore, the aspect ratios of transistors M2, M3 and bias current, set by voltage V_b, should be optimized. In order to maintain proper circuit operation, the minimum input swing range is set by the condition that transistors M1 and M4 work in their specified regions, which is given by $V_{in,min} = V_{gs,sf} + V_{ds,m3}$, where $V_{gs,sf}$ is the gate-source voltage of the transistor M2 and $V_{ds,m3}$ is the drain-source voltage of the transistor M3. It should be noted that owing to transistor M4 working in the weak inversion region, the input swing limit is caused by the correct operation of the source follower, and the V_{ds} of transistor M3 should be taken into consideration. On the other hand, when the operation region of transistor M4 is maintained, the maximum input voltage $V_{in,max}$ is limited due to the low supply voltage, which is given by $V_{in,max} = V_{thn} + V_{DD} - V_{Load}$, where V_{Load} is the voltage drop at loading transistors.

The linearity performance of the input signal will also be degraded by the non-ideal effect of the source follower. The nonlinearity term of NMOS source followers is dominated by the body effect [33]. The 180 nm mixed-signal CMOS process is a deep N-WELL process, so the body of the NMOS source follower can be connected to the source in a P-WELL to eliminate the body effect. However, mobility degradation and channel length modulation can also degrade the linearity performance of the source follower. The second-order distortion term of the source follower will be the dominant factor to affect the third-order harmonic distortion term in the drain current of weak inversion transistors. In addition, such a level shifter adds a high frequency pole of little influence.

2.4.2.2 Proposed Transconductor Implementation

Figure 2.24 shows the proposed transconductor circuit. The optimal ratios between the two voltage-to-current converters have been established to achieve minimal harmonic distortion. In the circuit, the transistor MR connected between two current mirrors would be working in the linear region as a resistor. Therefore, the continuous tuning of transconductance can be employed by adjusting the gate voltage, V_{tune}. The MOS resistor, which acts as the attenuator, actually promotes the linearity performance. In addition, it not only improves the linearity performance, but also adds a tuning strategy. Besides, the current mirror would introduce another high frequency pole, which is located at a lower frequency than that caused by the source follower. The non-dominant poles would only slightly influence the excess phase in this approach. Fortunately, the speed limitation caused by current mirror circuit would be relaxed by the addition of a small MOS resistor MR.

Figure 2.25 illustrates the large signal simulation of the proposed circuit. We can find that if no weak inversion transistors are used, the linearity performance is limited. When we increase the transconductance value of weak inversion transistors, high linearity performance can be obtained. However, if a larger transconductance value of weak inversion transistors is used, the third-order harmonic distortion term would be dominated by the weak inversion transistors. Besides, simulation also indicates a unity-gain bandwidth of 200 MHz with a 2pF loading and the phase margin of 89° is obtained.

In the pseudo-differential structure, in order to maintain low voltage operation, the use of stacked devices, such as the cascode structure, is prohibited. Therefore, to maintain the high output resistance in the pseudo-differential structure under the sub-micron process, increasing channel length could be one solution. However, a very long channel transistor will increase the parasitic capacitance associated with the output node, and thereby reduce the transconductor bandwidth. In Fig. 2.24, as the aspect ratio of transistor M11 equals to that of M12 and M13 equals to M14, the saturated transistors give the following expressions for both the differential-mode resistance and common-mode resistance.

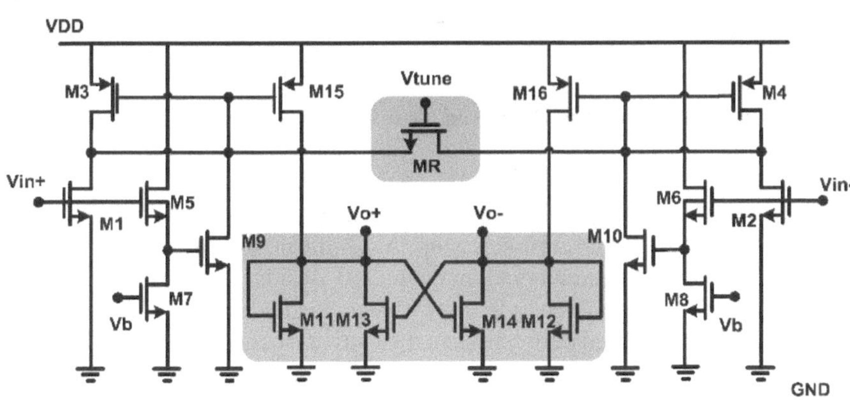

Fig. 2.24 Proposed OTA circuit

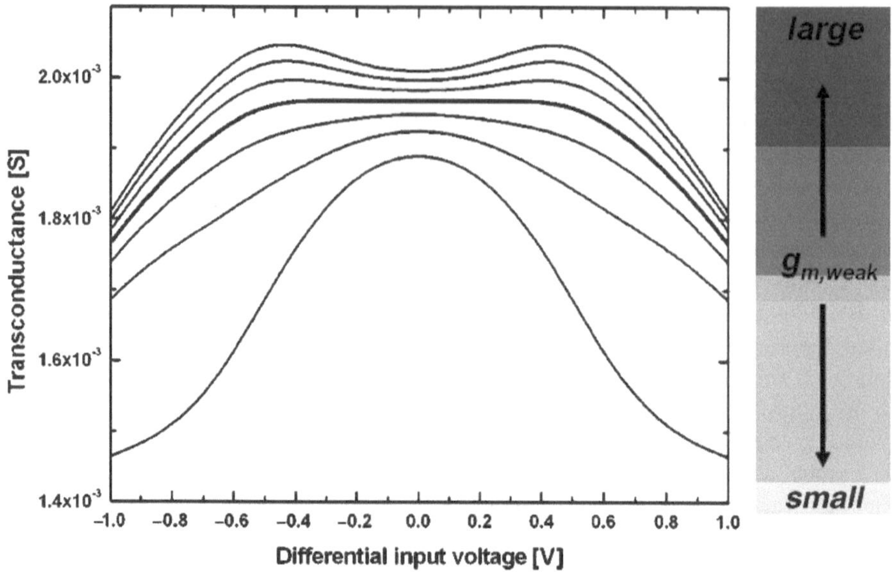

Fig. 2.25 Large signal simulation of the proposed circuit

$$\begin{aligned}\text{Common-mode resistance} &= (g_{m11} + g_{m13})^{-1} \\ \text{Differential-mode resistance} &= (g_{m11} - g_{m13} + 1/r_{o,all})^{-1}\end{aligned} \quad (2.42)$$

where gm_{11} and gm_{13} are the small signal transconductance of M11 and M13, respectively, and $r_{o,\ all}$ is the output resistance paralleled by the drain-source resistances of M11, M12, M13, M14, M15, and M16. By designing a little larger aspect value for M13 than M11, the negative conductance formed by $gm_{11} - gm_{13}$ would be close to the value of $1/r_{o,\ all}$ so as to increase the overall differential output resistance. The differential gain can be designed to a large value which will be suitable for most of applications. Moreover, as we can obtain the common-mode resistance from (2.42), the common-mode stability would be guaranteed by providing larger aspect ratios for transistors M11 and M13 than those for input transistors.

2.4.2.3 Nonidealities in the Proposed Transconductor

Mismatch between the input transistors of the transconductor will cause an imbalance in the drain current of the current mirrors, and thus generate even-order harmonics in voltage-to-current conversion. A significant increase in the device size will reduce the offset caused by random mismatch. Besides, the small MOS resistor helps to reduce the even-order harmonics by balancing the voltage at both terminals, and thus minimize the drain current mismatch caused by current mirrors. Therefore, with the MOS resistor, the dc offset caused by the device mismatch is reduced by allowing a different load resistance for the input transistors. For the output stage, mismatch between transistors would limit the differential mode gain. Careful layout was taken where the device match is required.

The output noise current of transconductor is the combination of saturated and weak inversion input transistors, current mirrors, and the MOS resistor. In a high-speed circuit design, the most significant noise source of a single transistor is the thermal noise generated in the channel. The channel noise can be modeled by a current source connected between the drain and source with a spectral density

$$\overline{I_n^2} = 4kT\delta g_{ms} \tag{2.43}$$

where k is the Boltzmann constant, g_{ms} is the source conductance, and the device noise parameter δ depends on the bias condition [25]. For the proposed circuit of the differential structure, we have defined $g_{m(n)} = g_{m(n+1)}$, where n equals to the odd number (ex: $g_{m1} = g_{m2}, g_{m3} = g_{m4}, \ldots$). Thus, the thermal noise density evaluated at the output node is derived as

$$\begin{aligned}\overline{I_{n,out}^2} = 8kT & \left\{ \left[\delta_1 \left(g_{m5} + g_{m7}\right) \left(\frac{g_{m9}}{g_{m5}}\right)^2 + \delta_1 \left(g_{m1} + g_{m3}\right) + \delta_2 g_{m9} \right] \right. \\ & \left. \times \left(\frac{g_{m3} R_{mr}}{2 + g_{m3} R_{mr}}\right)^2 + \frac{\delta_3}{2R_{mr}} \left(\frac{2}{2 + g_{m3} R_{mr}}\right)^2 \right\} \left(\frac{g_{m15}}{g_{m3}}\right)^2 \\ & + 8kT\delta_1 g_{m15}\end{aligned} \tag{2.44}$$

where δ_1, δ_2, and δ_3 are the noise parameter at saturation, weak inversion, and linear regions, respectively. Besides, R_{mr} is the resistance of the linear region transistor MR.

From the noise analysis, the source follower adds the input-referred noise while providing a voltage gain less than unity. Also, to reduce the thermal noise, the transconductance of input transistors should be maximized, which implies that an increasing overdrive input common-mode voltage $V_{ov} = V_{gs} - V_{thn}$ would result in a higher Signal-to-Noise Ratio (SNR).

2.4.3 Experimental Results

The chip was fabricated in 180 nm Deep N-WELL CMOS process. The micrograph of the chip is shown in Fig. 2.26 where the active area is 4.5×10^{-3} mm^2. The device size of saturated transistors M1 and M2, weak inversion transistors M9 and M10, and the linear transistor MR are 6 μm/0.18 μm, 1 μm/0.18 μm, and 15 μm/0.18 μm, respectively, and the bias voltage of V_b is 0.6 V. A supply voltage of 1-V was employed in the measurements and the nominal static power consumption of the transconductor is 2.5 mW. The output current versus input voltage over the tuning range is shown in Fig. 2.27. The measured harmonic distortion for a 400 mV$_{pp}$

Fig. 2.26 Die microphotograph

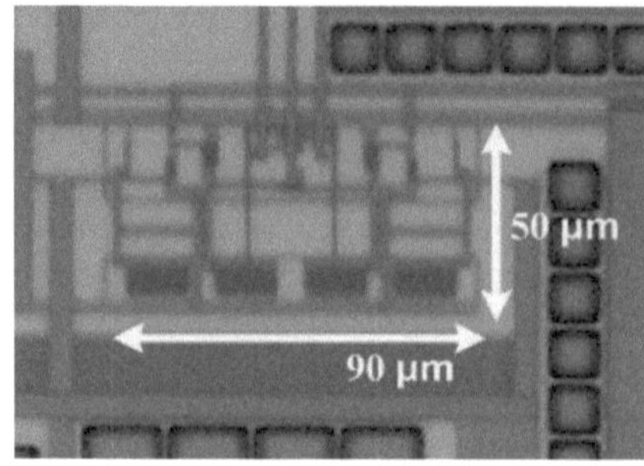

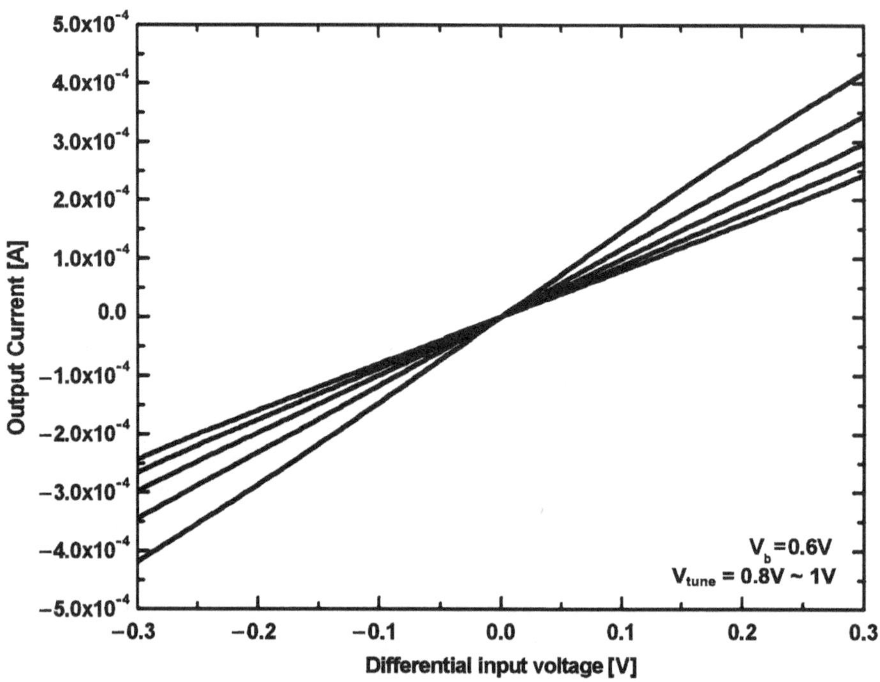

Fig. 2.27 The output current versus input voltage over the tuning range

differential input signal at 1 MHz is −70 dB, and the third-order harmonic distortion of the output current dominates the nonlinearity performance. The second-order harmonic distortion is due to the mismatch in the off-chip single-ended to differential input and differential output to single-ended conversion setup. The limitation of the input swing range is due to the departure of correct working regions under low supply voltage as predicted theoretically. The third-order inter-modulation distortion measured with two sinusoidal tones of 400 mV_{pp} amplitude is shown in Fig. 2.28. The IM3 is shown to be less than −52 dB at the speed of 50 MHz and drops to −50 dB at 56 MHz. The measured input referred noise spectral density at 50 MHz is 13 $nV/\sqrt{Hz}$. Table 2.2 summarizes the performance of this work with recently reported works.

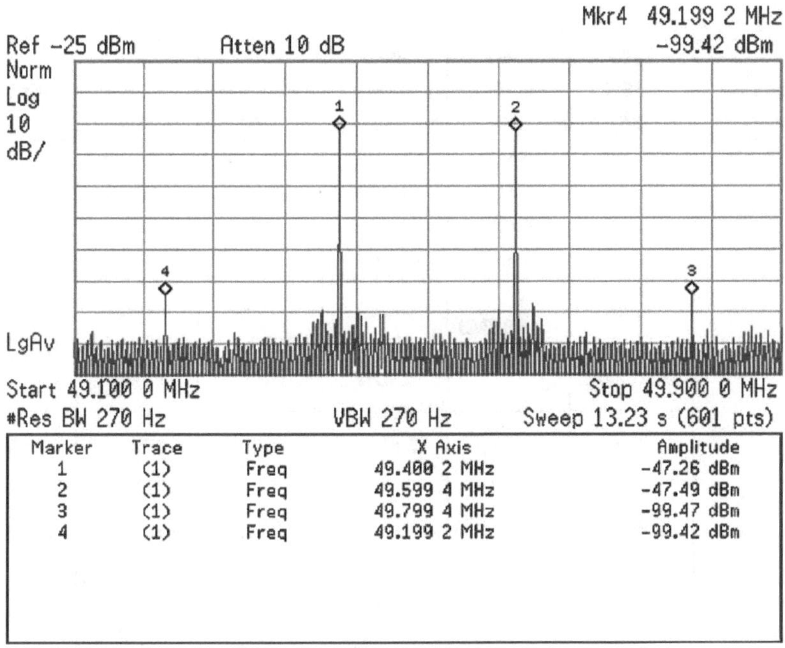

Fig. 2.28 Measured two tone inter-modulation distortions

2.4.4 Summary

A novel pseudo-differential transconductor based on a mobility compensation technique has been fabricated and measured. It is based on the principle that the third-order harmonic distortion term could be cancelled by the addition of the two drain current, one in the saturation region and the other in the weak inversion region, applying small extra power consumption by adding the linearity enhancement stage. The technique employed leads to a significant increase in linearity performance during voltage-to-current conversion. The measurement results show less than −52 dB IM3 at 50 MHz input signal under 1-V supply. We can conclude that the low-voltage transconductor could be provided as a high linearity and high-speed building block in analog VLSI applications.

2.5 Linear CMOS Transconductor in Nano-Scale CMOS Technology

In this section, high linearity MOSFET-only transconductor based on differential structures is presented. While a precise BSIM4 transistor model is introduced through analysis, the linearity can be improved by mobility compensation techniques as the device size is scaled down to the nano-scale CMOS technology.

Table 2.2 Comparison of previously reported works

Reference	2003 JSSC [26]	2004 TCAS-II [28]	2000 JSSC [34]	2006 TCAS-II [35]	2006 TCAS-II [36]	This work
Technology	0.5-μm CMOS (Measurement)	0.35-μm CMOS (Measurement)	0.6-μm CMOS (Measurement)	0.18-μm CMOS (Simulation)	0.35-μm CMOS (Simulation)	0.18-μm CMOS (Measurement)
HD3/IM3	-43dB HD3 at 30MHz	-65dB IM3 at 20MHz	-40dB HD3 at 5MHz	-65dB HD3 at 1MHz	-41.8dB HD3 at 1MHz	-55dB HD3 and -52dB IM3 at 50MHz
Input swing range	0.9Vpp	1.3Vpp	2.5Vpp	0.6Vpp	1 Vpp	0.4Vpp
Transconductance value	1065 μS	100 μS	105 μS	20 μS	30 μS	1000 μS
Supply	3.3V	3.3V	5V	1.8V	2.5V	1V
Power consumption	10.7mW	10.5mW	5.7mW	145 μW	150.8 μW	2.5mW
Normalized power efficiency (g_m/power)	0.25	0.024	0.04	0.34	0.5	1
Input referred noise	9.8 nV/√Hz	75 nV/√Hz	38 nV/√Hz	-	-	13 nV/√Hz

The measured results show an 18 dB improvement in the proposed version, and −65 dB HD3 can be achieved for a 2.1 MHz700 mV_{pp} differential input. Moreover, when the compensation devices use weak inversion transistors than the saturated transistors, the transconductance value can also be maintained implying higher power efficiency. The static power consumption is 130 μW under 1-V supply. This work is suitable as a building block for ADSL2 + applications, and the measurement results are in good agreement with theoretical analyses.

2.5.1 Introduction

In recent years, numerous linearization techniques have been designed and reported. The passive resistor, which is often implemented by the poly-silicon in the CMOS process, would be used in the high linearity transconductor circuit. Unfortunately, the use of passive resistors would cost more area, and it would also be affected by fabrication and temperature variation. Although the lack of the electronic tuning ability of passive resistor can be solved by using the resistor array, it encounters the accuracy problem while costing excessive areas and parasitic capacitors. Therefore, the transconductors with MOSFET-only configuration was developed and reported. In such a configuration, the linearity of such architecture is limited to below 50 dB owing to the non-ideal characteristic of the active device. Moreover, the degradation of linearity happens for small feature sizes of MOS transistors due to the influence of second-order effects, like velocity saturation and mobility reduction, under nano-scale CMOS technology.

In this section, the design of a low distortion and low supply voltage transconductor is presented. The proposed high linearity transconductor by using mobility compensation technique in the differential structure is discussed in Section 2.5.2. In Section 2.5.3, the measurement results of the proposed transconductor are discussed. The summary is concluded in Section 2.5.4.

2.5.2 *Operational Transconductance Amplifier*

2.5.2.1 The Transconductor in Differential Structure

As shown in Fig. 2.29, the simple operational amplifiers are formed by transistors M1 to M2, M3 to M4, and voltage shifter circuits. On one hand, transistors M7 and M8 operate in the linear region, and thus the input voltage variation can be passed down to the source of transistors M5 and M6, and then converted by transistors M7 and M8 for one voltage-to-current conversion path. On the other hand, the input voltage applied to the gate of the saturated transistors M9 and M10 composes another voltage-to-current conversion path. We should note that the Vsf voltage is composed by a simple source follower. Therefore, the linear relationship is given by

$$I_{out} = \frac{K_{lin}(V_{GS} - V_{thn})}{1 + \theta(V_{GS} - V_{thn})} V_{in} \tag{2.45}$$

where θ is the mobility reduction coefficient. The equation shows that the linear voltage-to-current conversion could be obtained by taking short channel effects into consideration. Unfortunately, high order nonlinearity components still occur in the V-I conversion especially for our nano-scale technology because Equation (2.45) is obtained from the analysis of an approximation.

2.5.2.2 The Modified Transconductor by Usinig Saturated Transistors

Figure 2.30 shows the modified transistor by using saturated transistors. To see how the circuit works, we should first deal with the drain current of the linear region transistor in a more detailed fashion. A precise LEVEL 54 BSIM4 transistor model can be expressed as [53]

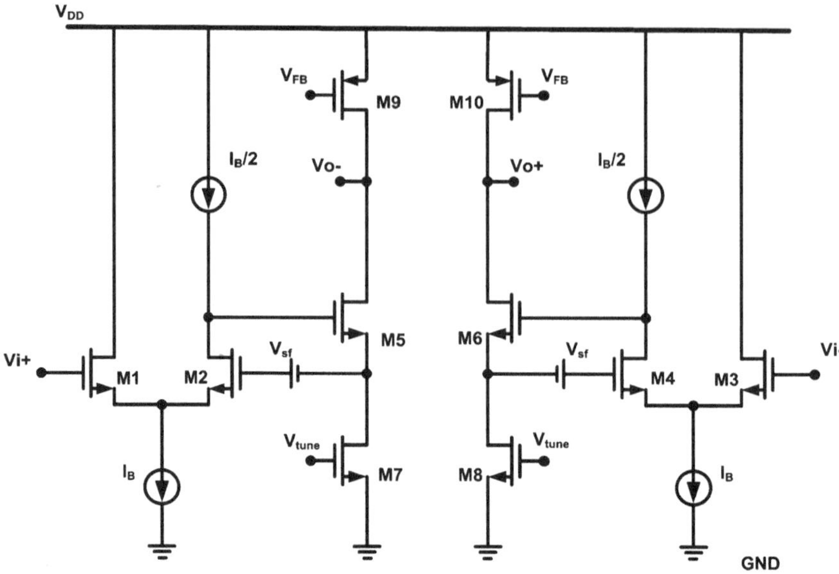

Fig. 2.29 The differential transconductor

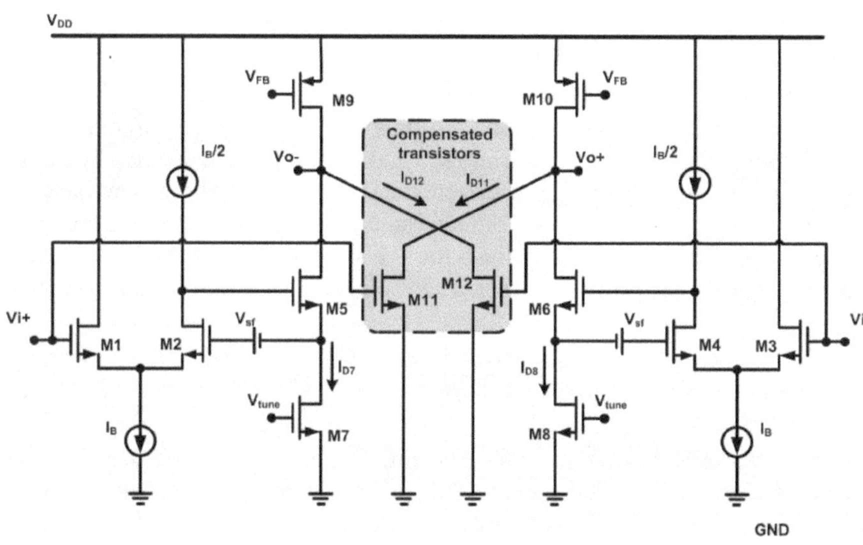

Fig. 2.30 The modified transconductor by using saturated transistors

$$
\begin{aligned}
I_D = & K_{lin}\left[\frac{(V_{GS}-V_{thn})\,V_{DS}-\frac{1}{2}V_{DS}^2}{1+\theta\,(V_{cm}-V_{thn})}\right]\times \\
& \left\{1+\frac{K_{lin}}{V_{DS}}\left[\frac{(V_{GS}-V_{thn})\,V_{DS}-\frac{1}{2}V_{DS}^2}{1+\theta\,(V_{cm}-V_{thn})}\right]\right\}^{-1}
\end{aligned}
\tag{2.46}
$$

By giving $V_{GS,7} = V_{GS,8} = V_{tune}$, $V_{DS,7} = V_{cm} + v_d/2$, and $V_{DS,8} = V_{cm} - v_d/2$, where V_{cm} is the input common-mode voltage and v_d is the differential voltage, we can derive the differential voltage-to-current characteristic of transistors M7 and M8. To analyze the linearity of the drain-to-source voltage against the drain current, a Taylor series expansion is used and the relationship can be expressed as

$$\begin{aligned} I_{D,lin} &= I_{D7} - I_{D8} \\ &= a_{1,lin}v_d + a_{2,lin}v_d^2 + a_{3,lin}v_d^3 + a_{4,lin}v_d^4 + \cdots \end{aligned} \tag{2.47}$$

where

$$a_{1,lin} = \frac{K_{lin}\left\{K_{lin}V_{cm}^2 + 4\left(V_{tune} - V_{th} - V_{cm}\right)\left[1 + \left(K_{lin} + \theta\right)\left(V_{tune} - V_{th}\right)\right]\right\}}{\left[-2 + K_{lin}(V_{cm} - 2V_{tune} - V_{th}) - 2(V_{tune} - V_{th})\theta\right]^2}$$

$$a_{3,lin} = \frac{-K_{lin}^2\left[1 + \theta\left(V_{tune} - V_{th}\right)\right]\left[1 + \left(K_{lin} + \theta\right)\left(V_{tune} - V_{th}\right)\right]}{\left[-2 + K_{lin}(V_{cm} - 2V_{tune} - V_{th}) - 2(V_{tune} - V_{th})\theta\right]^4}$$

In the equation, the even-order harmonic term can be cancelled out by the differential structure and thus the third-order harmonic distortion would become the dominant component.

In this modified transconductor, the other saturated transistors based on the differential structure are used to compensate the third-order harmonic distortion. The concept of topology comes from the fact that the output current of the saturated transconductor in the differential pair also suffers the problems of channel length modulation and short channel effects as the feature size of the transistors is chosen to be small, and thus the output current is given by

$$I_{D,sat} = \frac{K_{sat}\left(V_{GS} - V_{th}\right)^2}{2\left[1 + \theta\left(V_{GS} - V_{th}\right)\right]} \tag{2.48}$$

where K_{sat} is the device parameter of saturated transistors. By giving $V_{GS,11} = V_{cm} + v_d/2$, and $V_{GS,12} = V_{cm} - v_d/2$, a Taylor series expansion is introduced and then the voltage-to-current relationship can be expressed as

$$\begin{aligned} I_{D,sat} &= I_{D11} - I_{D12} \\ &= a_{1,sat}v_d + a_{2,sat}v_d^2 + a_{3,sat}v_d^3 + a_{4,sat}v_d^4 + \cdots \end{aligned} \tag{2.49}$$

where

$$a_{1,sat} = \frac{K_{sat}\left(V_{cm} - V_{th}\right)V_{thn}\left[2 + \theta\left(V_{cm} - V_{th}\right)\right]}{\left[1 + \theta\left(V_{cm} - V_{th}\right)\right]^2}$$

$$a_{3,sat} = \frac{-K_{sat}\theta}{\left[1 + \theta\left(V_{cm} - V_{th}\right)\right]^4}$$

We can find that the dominant distortion of the saturated transistor occurs due to the short channel effects in the third-order harmonic component under the topology of differential structure. The introduced transistor in the saturation region seems to provide another distortion term. However, although the third-order harmonic terms of Equations (2.47) and (2.49) have the same signs, we can put the output node of saturated transistors in the opposite position corresponding to the linear region transistors. Thus, when fixed values of parameters are given, the aspect ratio of the compensated transistor can be found to achieve a highly linear voltage-to-current conversion. However, the main disadvantage of the modified circuit is the reduced transconductance.

2.5.2.3 The Modified Transconductor by Using Weak Inversion Transistors

In order to keep the linearity performance with the maintained transconductance, the weak inversion transistors are used to replace the saturated transistors as shown in Fig. 2.31. By giving a voltage shifter to make sure the weak inversion operation, the drain current of the weak inversion transistors can be expressed as

$$I_{D,sub} = I_0 \left(\frac{W_w}{L_w}\right) e^{\frac{V_{GS}}{\xi V_T}} \left(1 - e^{\frac{-V_{DS}}{V_T}}\right) \tag{2.50}$$

where W_w and L_w are the width and length of the device, respectively, I_o is the process-dependent parameter, ζ is the weak inversion slope factor, and V_T is the thermal voltage. Again, a Taylor series expansion is introduced while the voltage follower is well maintained, and then the voltage-to-current relationship can be expressed as

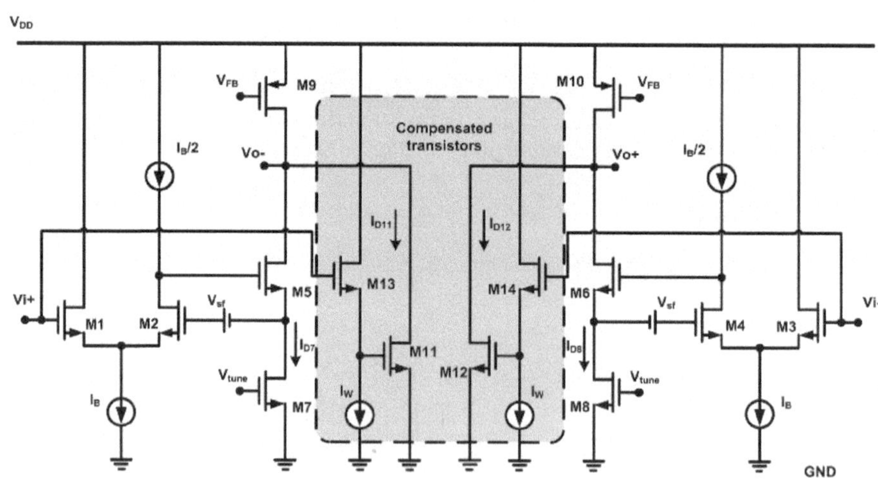

Fig. 2.31 The modified transconductor by using weak inversion transistors

$$\begin{aligned} I_{D,sub} &= I_{D11} - I_{D12} \\ &= a_{1,sub} v_d + a_{2,sub} v_d^2 + a_{3,sub} v_d^3 + a_{4,sub} v_d^4 + \cdots \end{aligned} \tag{2.51}$$

where

$$a_{1,sub} = \frac{I_0}{\xi V_T}\left(\frac{W_w}{L_w}\right) e^{\frac{V_{cm}-V_{sf}}{\xi V_T}}$$

$$a_{3,\, sub} = \frac{I_0}{24\,(\xi V_T)^3}\left(\frac{W_w}{L_w}\right) e^{\frac{V_{cm}-V_{sf}}{\xi V_T}}$$

The third-order harmonic distortion term decreases with smaller device width and smaller input common-mode voltage. When the optimal sizes are designed, the combined current provides a higher transconductance with minimized third-order distortion term. The function of differential input voltage to transconductance variation and the predicted performance are shown in Fig. 2.32.

2.5.3 Experimental Results

The chip was fabricated by the TSMC 180 nm CMOS process. A supply voltage of 1-V was employed and the nominal static power consumption of the transconductor

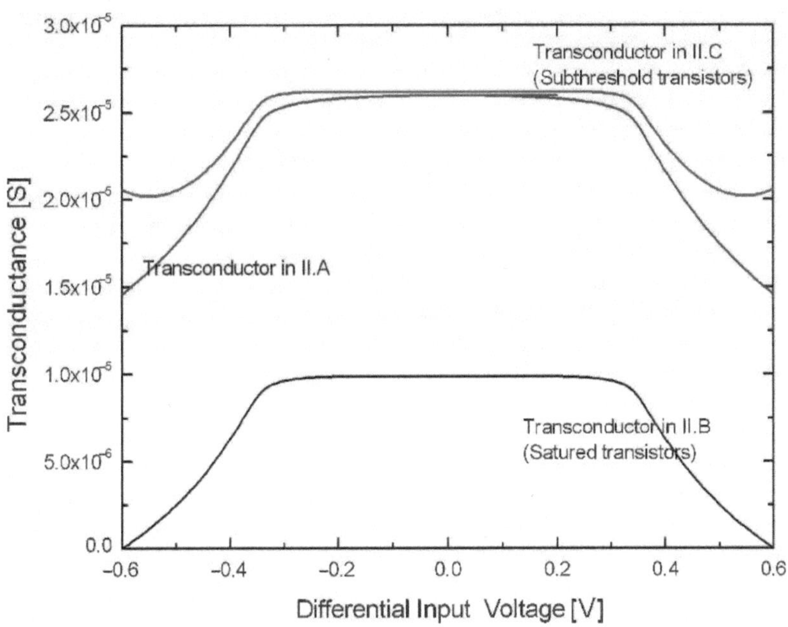

Fig. 2.32 The simulated Gm range of the proposed transconductor

is 130 μW. The chip micrograph is shown in Fig. 2.33 with less than 27.1×10^{-3} mm^2, and these transconductors almost use the same area. The HD3 measured with a sinusoidal tones of 0.7 V_{pp} amplitude at the speed of 2.1 MHz is shown in Fig. 2.34. Figure 2.34a shows the transconductor with no improvement. As shown in Fig. 2.34b, the HD3 is improved to about 16 dB, but the decrease of transconductance is 10 dB. Then, the HD3 in Fig. 2.34c is shown to be about −65 dB, and an 18 dB improvement when compared with Fig. 2.34a. We can find that the transconductance is almost the same. In addition, the output referred noise spectral density at 2.1 MHz is 52 nV/$\sqrt{\text{Hz}}$.

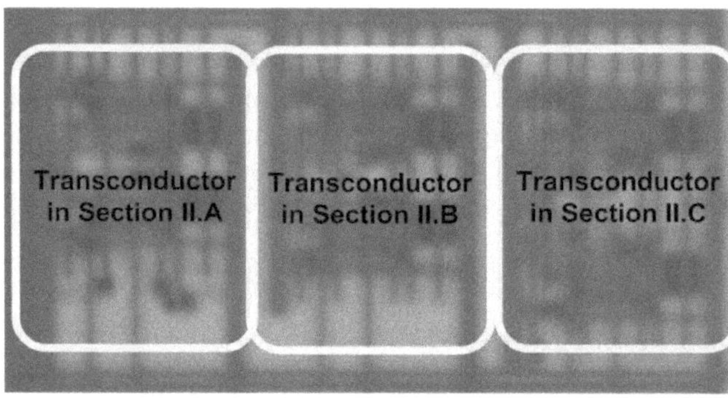

Fig. 2.33 The chip micrograph

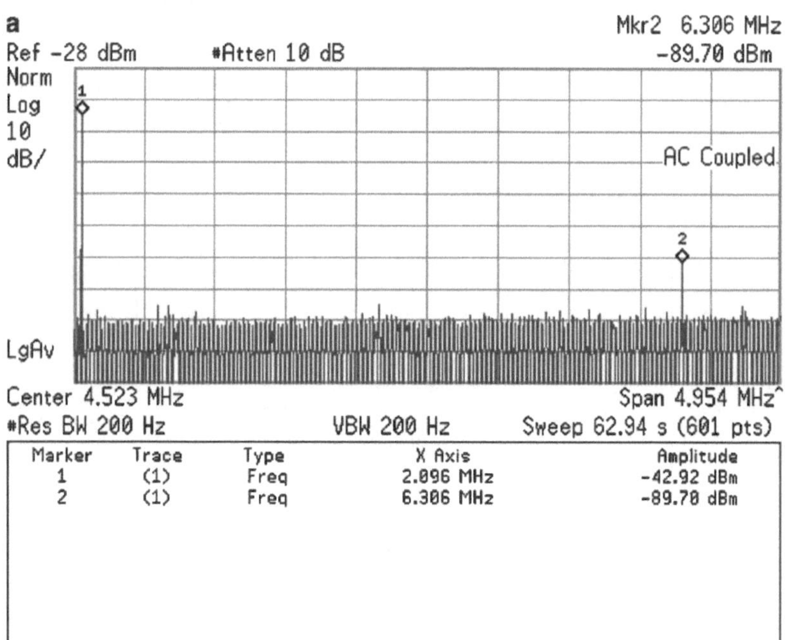

Fig. 2.34 Measured third-order harmonic distortion: (**a**) the differential transconductor

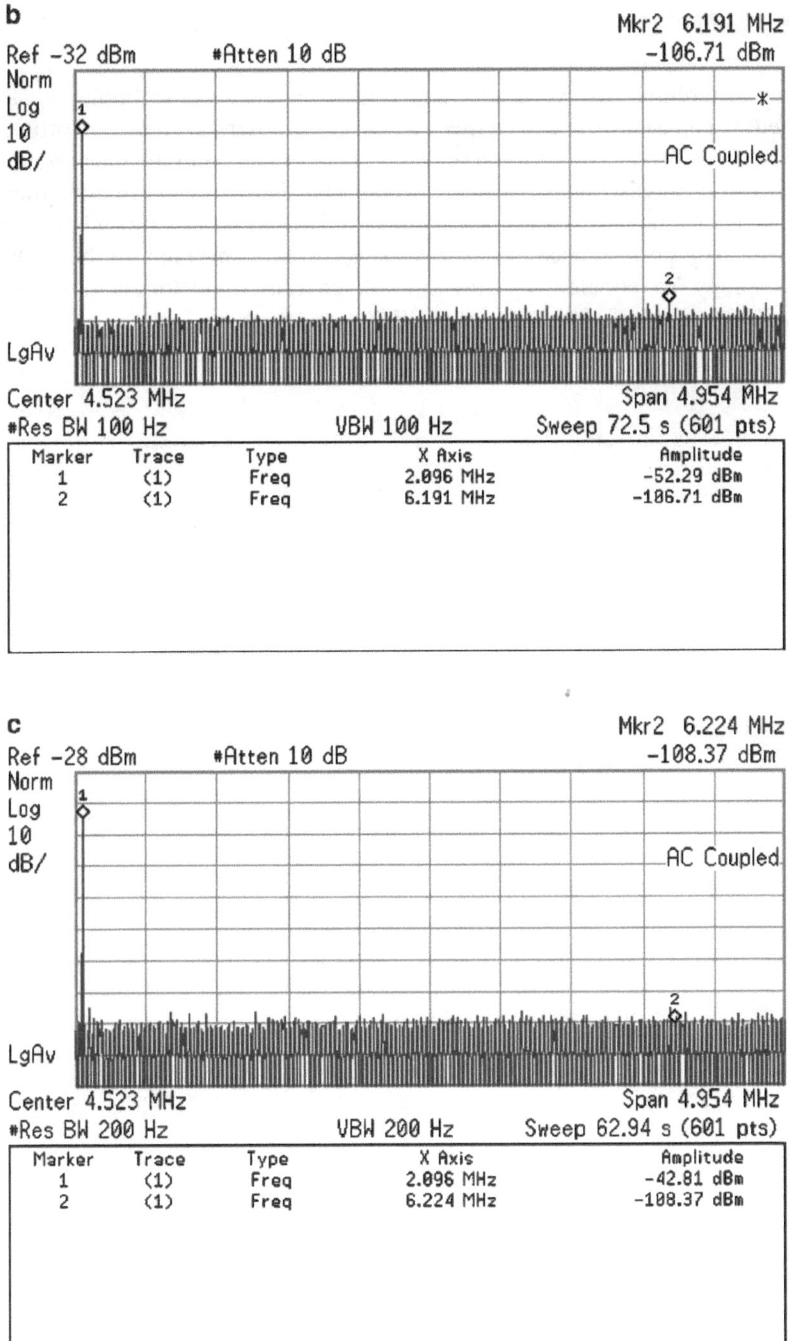

Fig. 2.34 (continued) (**b**) the modified transconductor by using saturated transistors (**c**) the modified transconductor by using weak inversion transistors

2.5.4 Summary

A novel differential transconductor under nano-scale CMOS technology has been reported. It is based on the principle that the third-order harmonic distortion term could be cancelled by the addition of two drain current, one in the linear region and the other in the weak inversion region, to improve the linearity while the transconductance is maintained. The technique employed leads to a significant improvement in the linearity performance in the voltage-to-current conversion in the MOSFET-only topology. We conclude that the low voltage transconductor could be provided as a high linearity building block in ADSL2 + applications.

Chapter 3
G_m-C Filter

3.1 Introduction

This chapter discusses the implementation of the transconductor. It is designed as a basic building block for the G_m-C analog filter. The basic concept of a G_m-C filter is discussed.

3.2 Implementation of the G_m-C Filter

This section shows the implementation of the G_m-C filter. First, the fundamental of the integrator is discussed. Then, the methods of filter synthesis are introduced. The transconductors and capacitors are assumed ideal under filter synthesis. After discussing the synthesis methods, the non-idealities of the filter are presented. A second-order band-pass filter is used as an example to illustrate the effects of the transconductor non-idealities. It will show that transconductor properties determine the filter performance.

3.2.1 Integrator

To realize an integrator in G_m-C technology, a transconductor and a capacitor can be used as shown in Fig. 3.1. We assume that the transconductor and the capacitor are ideal elements. From the figure, the output current is applied to the integrating capacitor, C_L, and the output voltage is given by

$$V_o = \frac{G_m}{SC_L} V_i \tag{3.1}$$

T.-Y. Lo and C.-C. Hung, *1V CMOS* G_m-C *Filters: Design and Applications*, Analog Circuits and Signal Processing,

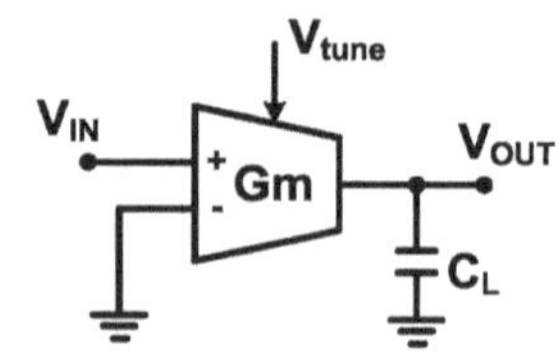

Fig. 3.1 The single-ended integrator

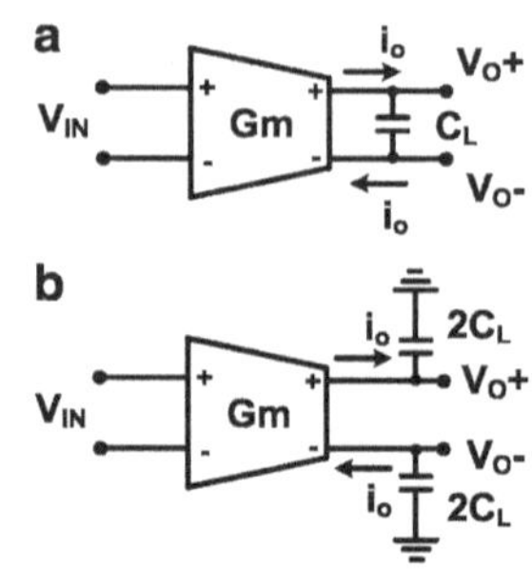

Fig. 3.2 The fully-differential integrator: (**a**) single capacitor (**b**) dual capacitors

We can obtain that the unity-gain frequency ω_o is G_m/C_L. Thus, the output voltage is equal to the integration of the input voltage multiplied by the integrator unity-gain frequency.

The ideal integrator has an infinite DC gain and a phase shift of $-90°$. We can model (3.1) to have the transfer function as $H(j\omega\omega) = (R(\omega) + jX(\omega))^{-1}$. The quality factor, which is defined as $Q(\omega) = X(\omega)/R(\omega)$, indicates the integrator phase deviation from $-90°$. If the integrator is ideal, the quality factor would be infinite.

In most of the integration circuits, it is required to keep the signals fully differential. The fully differential circuits show good noise and distortion properties as described in Chapter 2. The fully differential integrator can be realized by two structures, as shown in Fig. 3.2. The integrator in Fig. 3.2a requires one fourth capacitance than the ones in Fig. 3.2b for the same unity-gain frequency. However, the single capacitor in Fig. 3.2a would be realized by using integrated capacitors, and the structure consists of a significant amount of parasitic capacitors. For high frequency applications, the ground capacitors would be suitable owing to small loading capacitors. Besides, the top and bottom plate parasitic capacitors from the integrated capacitor would cause symmetry problems. Thus, two parallel capacitors, which have half of original capacitance, with opposite top-bottom direction are often realized.

The non-ideal transistor is characterized by a non-zero output conductance g_o and a delay τ_b in the transfer function. The output conductance is usually a positive small value. The delay would be due to the parasitic poles and zeros and they are located at higher frequencies. Since the parasitic zeros and poles are located at higher frequencies than the band of transfer function, these effects can be modeled with a single equivalent zero. The zero can be located in the right half-plane or left half-plane under s-domain to model phase lag or lead. For the integrator, the dominant pole obtained from the loading capacitor and the gain are given by

$$\tau_a = \frac{C_L}{g_o}, A_i = \frac{g_m}{g_o} \tag{3.2}$$

The transfer function of the integrator is given by

$$H_{int} = A_i \frac{1 - s\tau_b}{1 + s\tau_a} \tag{3.3}$$

Figure 3.3 shows the transfer function of the non-ideal integrator with respect to the ideal model. The phase lag is modeled as a right half-plane zero. The finite DC gain and the parasitic zero would cause the deviation of the integrator phase from −90°. The excess phase is defined at the integrator unity-gain frequency as

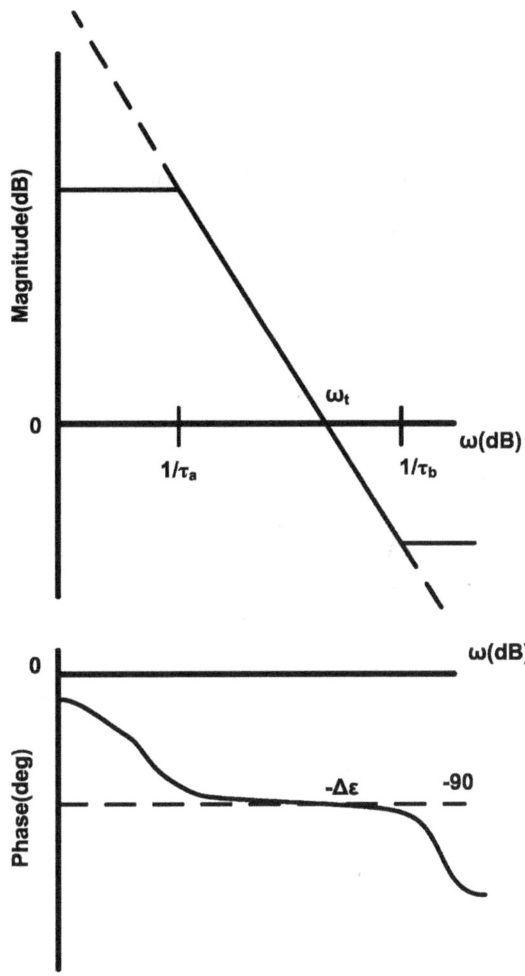

Fig. 3.3 The gain and phase of ideal (dashed) and non-ideal (solid) integrator

$$\Delta\varepsilon(\omega) = -\{90^\circ + \arg[H_{int}(j\omega)]\} \tag{3.4}$$

The large value of excess phase would be the major source of errors in the filter transfer function. From (3.3), the transfer function can also be modeled as

$$H_{int}(j\omega) = \frac{1}{R_{nonideal}(\omega) + jX_{nonideal}(\omega)} \tag{3.5}$$

The quality factor of the integrator is defined as

$$Q_{int}(\omega) = \frac{X_{nonideal}(\omega)}{R_{nonideal}(\omega)} = \tan(-\arg(H_{int}(j\omega))) \tag{3.6}$$

This equation also shows that the integrator can be expressed as the phase error for a small phase error. If we introduce the value in (3.3) and assume $\tau_b \ll \tau_a$, the integrator quality factor can be calculated as

$$\frac{1}{Q_{int}(\omega)} = \frac{1}{\omega\tau_a} - \omega\tau_b \tag{3.7}$$

From the above equation, the quality factor could be infinite at a specific condition. This condition will hold if the output conductance g_o and the delay τ_b have the same sign, and ω is the geometric mean of the dominant and parasitic pole. Thus, the phase is compensated from the cancellation.

3.2.2 *Programmable Integrator*

In the integrator design, the unity-gain frequency would be an important factor for future filter synthesis. However, the value varies under process fabrication, and an automatic tuning circuit is required. The main function of the automatic tuning circuit is discussed in Chapter 5. The unity-gain frequency is dependent on the transconductance and the loading capacitor. Adjustable transconductor or capacitor is required to compensate process corner variation. Thus, we can keep transconductance constant and vary the capacitance or vice versa. These approaches are called constant-G_m and constant-C_L, respectively. For these approaches, the performance of noise, area, and power dissipation is different.

Figure 3.4 shows the implementation of the programmable integrator. In Fig. 3.4a, the unity-gain frequency is defined by the largest transconductance. The output noise current power spectral density v_o^2 is defined to be $4\,kT\eta G_m$, where η is the excess noise factor of the transconductor. The power dissipation is assumed to be proportional to the transconductance with a factor b. For a given v_o^2, the value of C_L is defined and then we obtain a largest value, ω_{max}, through the largest transconductance.

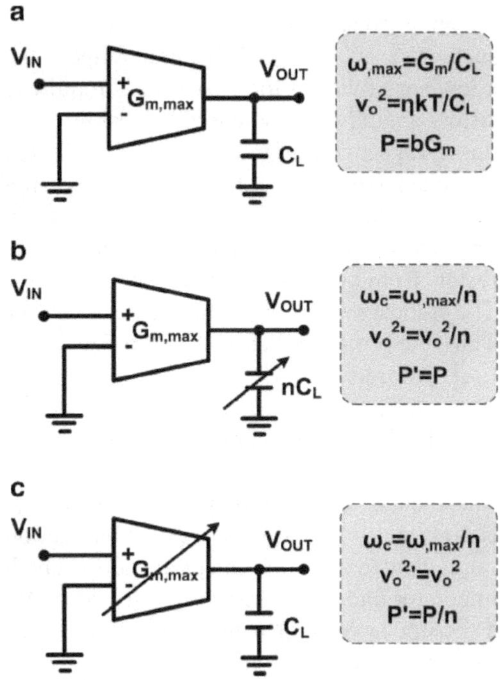

Fig. 3.4 Integrator: (**a**) the integrator with maximum unity-gain frequency (**b**) the integrator with constant-Gm design (**c**) the integrator with constant-C_L design

Figures 3.4b and c shows the corresponding value when the unity-gain frequency is adjusted. Thus, the noise, capacitor area, and the power will vary accordingly. For the constant-G_m design, a large capacitor area is required and it is not suitable for high frequency applications. The low noise performance is not very helpful since only the highest noise contribution is needed to be defined over the frequency tuning range. In contrast, the constant-C_L design decreases the transconductance, which implies lower power consumption. The noise performance is equal throughout the frequency tuning range without any overdesign. Therefore, the constant-C_L approach is preferred in our filter design.

In actuality, the excess noise factor will vary during frequency tuning. For the source degeneration transconductor, the factor increases by decreasing the transconductance. This condition implies that large effort of integrator noise should be taken at low frequency applications. In contrast, the accuracy of frequency response becomes important owing to the large excess phase at high frequency applications.

By using a linear time scaling technique [37], we can prove that the integrated noise performance is constant over the frequency tuning range while taking the excess noise factor and bandwidth into consideration.

3.2.3 Filter Synthesis Methods

In this section, two filter syntheses methods are presented. One is based on the biquad section and the other is the signal flow graph. Although different topologies are obtained at the circuit level through the syntheses methods, the same properties such as sensitivities and dynamic range should be maintained. At the circuit level, a low sensitivity of the passband is required, and thus the matching between transistors should be good enough. Besides, the optimal dynamic range is defined by the voltage swing of internal nodes, and we need to adjust the transconductance or capacitance in the loop for optimized dynamic range.

When starting with a passive prototype, the LC-ladder filter with the property of low sensitivities to component variation in the passband is chosen. Therefore, the results from synthesis show the same low sensitivities for the LC-ladder structure. The voltage swings of internal nodes at LC-ladder prototype are close to each other. Thus, the prototype has near optimized dynamic range.

3.2.3.1 Biquad Sections

The biquad synthesis is a method to realize the filter transfer function by cascading multiple first or second order sections. A biquad transfer function can be implemented by a two-integrator loop. The loop is composed by a lossy inverting integrator and a non-inverting integrator. An integrator is lossy when we connect a resistor in parallel with the loading capacitor. The resistor can be implemented by a transistor within negative feedback. Figure 3.5 shows the realization of a lossy integrator. The transfer function is given by

$$\frac{V_o}{V_i} = \frac{-g_{m1}}{sC_L + g_{m2}} \tag{3.8}$$

A general biquad section has a second order function given by

$$\frac{V_o}{V_i} = K\frac{a_0 + a_1 s + a_2 s^2}{s^2 + s\frac{\omega_o}{Q_p} + \omega_0^2} \tag{3.9}$$

where ω_o is the filter center frequency and Q_p is the filter pole quality factor.

The low-pass, high-pass, band-stop, band-pass, and all-pass functions are dependent on the constants a_0, a_1, and a_2. We should note that a_0 is equal to ω_0^2 and thus K is the DC gain of the transfer function. Figure 3.6 shows a G_m-C implementation of the biquad section. The transfer function from input V_i to the terminal V_B forms a band-pass response and is given by

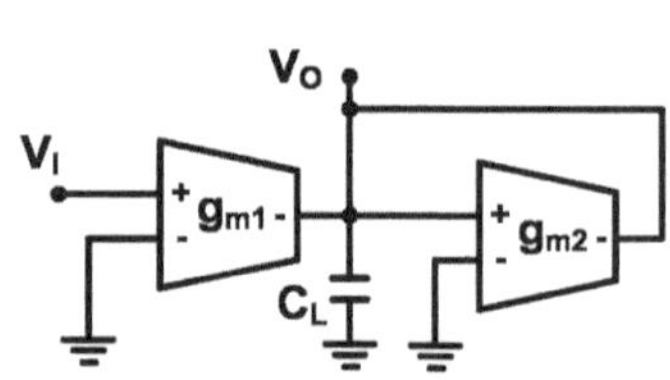

Fig. 3.5 The lossy integrator

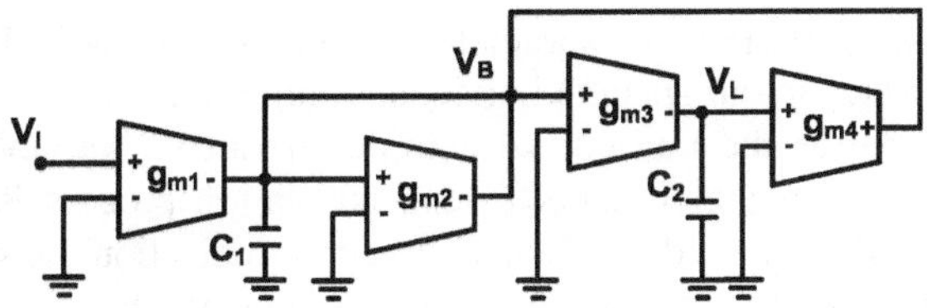

Fig. 3.6 The biquad section

$$\frac{V_B}{V_i} = \frac{s g_{m1} C_2}{s^2 C_1 C_2 + s C_2 g_{m2} + g_{m3} g_{m4}} \tag{3.10}$$

At terminal V_L, we can obtain a low-pass response

$$\frac{V_L}{V_i} = \frac{g_{m1} g_{m3}}{s^2 C_1 C_2 + s C_2 g_{m2} + g_{m3} g_{m4}} \tag{3.11}$$

Comparing (3.9), the cutoff frequency ω_0 and the quality factor Q can be expressed as

$$\omega_o = \sqrt{\frac{g_{m3} g_{m4}}{C_1 C_2}} \tag{3.12}$$

$$Q_p = \sqrt{\frac{g_{m3} g_{m4} C_1}{g_{m2}^2 C_2}} \tag{3.13}$$

$$K = \frac{g_{m1}}{g_{m4}} \tag{3.14}$$

To simplify a low-pass transfer function, we can set the value of g_{m1}, g_{m2}, and g_{m4} to be the same. This will set the gain of the transfer function to be 1. Also, the capacitors C_1 and C_2 can be set to have the same value. Thus, we can obtain $\omega_0 = g_{m1}/C_1$ and $Q_p = g_{m1}/g_{m2}$.

The advantage of the biquad synthesis method is the cascade fashion, and thus the loop is very stable even at higher order transfer function. Moreover, each node of the biquad has a capacitance to ground, and it is suited for high frequency operations. The main disadvantage is the loading effect caused by cascading. The effect can be minimized by calculating the input capacitance of the next stage and then taking the value into consideration. Besides, the sensitivity of the biquad section to component variation is larger than the LC-ladder structure.

3.2.3.2 Signal Flow Graph

Before we discuss the signal flow graph synthesis, the denormalization of the frequency and impedance of the passive component is introduced at first. Since the passive prototype of the transfer parameters is the normalized value, we need to denormalize the value to our specific impedance and frequency.

If we consider p as the normalized complex-frequency variable and s as the denormalized one, then the frequency denormalization process is defined to be $s = \Omega_n p$, where Ω_n is called the frequency-denormalization constant. When we consider an inductor with a frequency normalized value of L henrys, the corresponding frequency normalized impedance $Z_n(p) = pL$. The denormalized impedance would be $Z(s) = sL/\Omega_n$, and thus it represent an inductor of value L/Ω_n. Similarly, a denormalized capacitor would have a value of C/Ω_n. In addition, the resistor would keep the same value because the complex-frequency variable is independent on the impedance. The impedance denormalization could be obtained by the relation $Z(s) = z_n Z_n$, where z_n is called the impedance-denormalization constant. Again, we consider an inductor with an impedance normalized value of L henrys. The impedance-normalized impedance is $Z_n = L$, and the denormalized impedance is $Z = z_n L$, which represents an inductor of value $z_n L$. On the other hand, the impedance-normalized impedance for a capacitor with the value of C farads is $Z_n = 1/C$, and the denormalized impedance is $Z = z_n/C$. Thus, the capacitance becomes C/z_n. We can find that the inductance and capacitance move to opposite directions under impedance denormalization, rather the same direction of the frequency denormalization.

The signal flow graph converts the passive diagram to a symbolic diagram, and it shows the relation of node voltage and branch current. Through the method, the transconductors would be identical, and then the scaled element value achieves maximum dynamic range. Moreover, the number of integrators should be equal to the filter order after the synthesis.

Figure 3.7 shows a passive diagram of sixth-order Elliptic low-pass filter. We divide the filter into three parts: the input R network, the LC ladder network, and the output RC network.

Firstly, we discuss the LC ladder network by using signal flow graph synthesis. The typical LC ladder with defined voltage at nodes and current at branches is shown in Fig. 3.8a. One of the current equations can be given by

$$I_{l1} - I_{l3} = (V_1 - sL_1 I_{l1} - V_3)\frac{1}{sL_2} \tag{3.15}$$

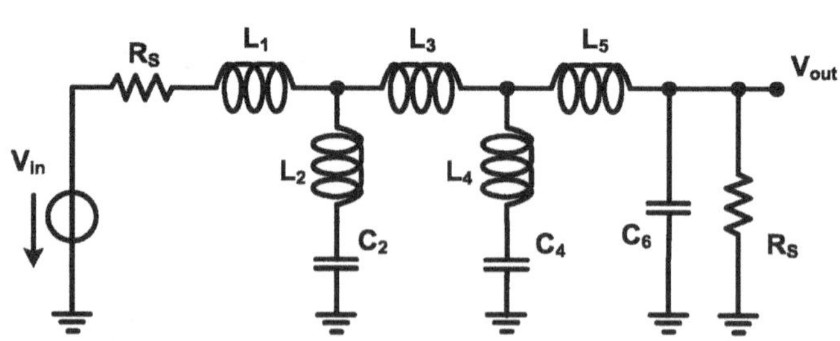

Fig. 3.7 The sixth-order Elliptic low-pass passive filter

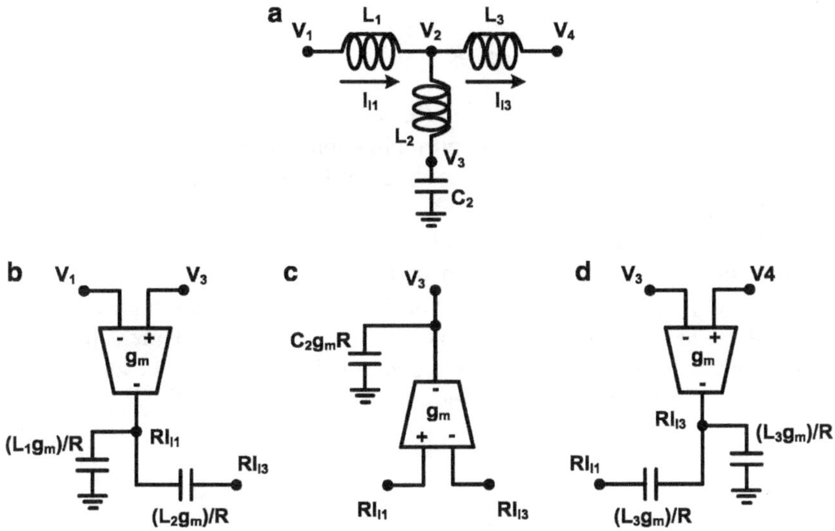

Fig. 3.8 The SFG synthesis of the LC network; (**a**) passive prototype (**b**) implementation of (3.16) (**c**) implementation of (3.17) (**d**) implementation of (3.18)

Upon multiplying the above equation by a normalization resistor R and a transconductance g_m, we have

$$R\left(I_{l1}-I_{l3}\right)=\left[\left(V_1-V_3\right)g_m-\frac{sL_1g_m}{R}\left(RI_{l1}\right)\right]\frac{R}{sL_2g_m} \tag{3.16}$$

Then, the circuit realized from (3.16) is shown in Fig. 3.8b. We can also obtain the following equations from the current equations of Fig. 3.8a

$$V_3=\left[\left(RI_{l3}-RI_{l1}\right)g_m\right]\frac{1}{sC_2Rg_m} \tag{3.17}$$

$$RI_{l3}=\left[\left(V_3-V_4\right)g_m-\frac{sL_2g_m}{R}\left(RI_{l3}-RI_{l1}\right)\right]\frac{R}{sL_3g_m} \tag{3.18}$$

The circuits realized from (3.17) and (3.18) are shown in Fig. 3.8c and Fig. 3.8d, respectively.

For the input R and output RC network in Fig. 3.7, we set $R_s = 1/g_m$ and thus the impedance-denormalization constant z_n is equal to $1/g_m$. In this condition, the normalization resistor R is equal to $1/g_m$. Then, the equations obtained from the passive network are given by

$$g_m\left(V_{in}-V_1\right)=g_m\frac{R_s}{R}\left(RI_{l1}\right) \tag{3.19}$$

$$V_o = \frac{1}{sC_6 g_m R}\left(RI_{l5} - \frac{RV_o}{R_s}\right) g_m \tag{3.20}$$

The circuits realized from (3.19) and (3.20) are shown in Fig. 3.9a and Fig. 3.9b, respectively. Since the impedance denormalization, $R_s = 1/g_m$, is introduced, the value of capacitance and inductance should be denormalized as well. After synthesizing the sixth-order passive network, some transconductor will share the same nodes and we can combine them together to reduce active elements. Figure 3.10 shows the final active implementation of the sixth-order Elliptic low-pass filter based on the G_m-C structure. The value of the equivalent capacitor should be carefully transformed. For a given transconductance and passive element parameters, the equivalent capacitor results from the combination of impedance and frequency denormalization after signal flow synthesis.

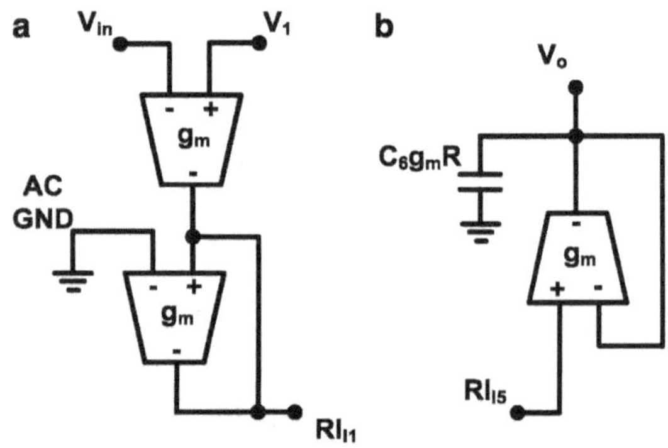

Fig. 3.9 The SFG synthesis of the input R and output RC network; (**a**) equation (3.19) (**b**) equation (3.20)

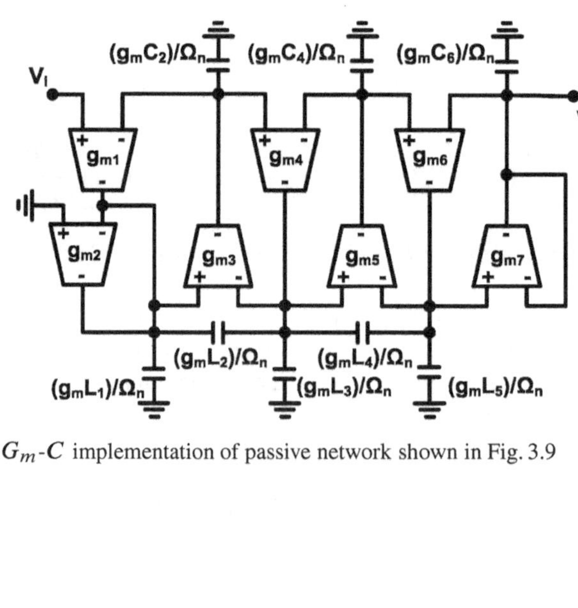

Fig. 3.10 The G_m-C implementation of passive network shown in Fig. 3.9

3.2.4 *Effect of Integrator Non-Idealities in Filter*

The G_m-C filter usually consists of more than one integrator. The non-idealities would be modeled by using the non-ideal integrator discussed in Section 3.2.1. Since the performance of the filter is highly dependent on the transconductor, the finite integrator quality factor should be taken into account. Therefore, the transconductance in the filter would need to add an effective zero by taking parasitic effects. Besides, the finite transconductor output conductance is connected parallel to the integrating node, and the loss of the loading capacitance is caused by the sum of non-zero conductance. Thus, the transconductance and loading effect caused by non-idealities could be modeled as

$$g_{m,nonideal} = g_m\,(1 - s\tau_b) \tag{3.21}$$

$$C_{L,nonideal} = C_L + \sum g_{oi} \tag{3.22}$$

To discuss the filter non-idealities, the biquad circuit shown in Fig. 3.6 with the band-pass function is illustrated as an example. The ideal transfer function is given by (3.10).

3.2.4.1 Non-Zero Output Conductance

When we take the effect of non-zero output conductances in (3.22), the loading capacitance C_1 and C_2 becomes

$$\begin{aligned} C_{1,go} &= C_1 + \frac{g_{oi1}}{s} \\ C_{2,go} &= C_2 + \frac{g_{oi2}}{s} \end{aligned} \tag{3.23}$$

where $go_{i1} = go_2 + go_3 + go_4$ and $go_{i2} = go_1$. If we substitute (3.23) into (3.10), the result is given by

$$H_{nonideal,go}(s) = \frac{s\frac{g_{m1}}{C_1} + \frac{g_{m1}g_{oi2}}{C_1C_2}}{s^2 + s\left(\frac{g_{m2}}{C_1} + \frac{g_{oi1}}{C_1} + \frac{g_{oi2}}{C_2}\right) + \frac{g_{m3}g_{m4} + g_{oi2}(g_{m2}+g_{oi1})}{C_1C_2}} \tag{3.24}$$

From the above equation, the gain of the transfer function will be slightly higher than the ideal value at low frequency. Also, we can obtain the affected center frequency $\omega_{o,go}$ and the affected quality factor $Q_{p,go}$. For a high Q implementation, we can assume $g_m =, g_{m3} = g_{m4} \gg g_{m1} = g_{m2}$, $g_o = g_{o3} = g_{o4} \gg g_{o1} = g_{o2}$ and $C = C_1 = C_2$, the center frequency can be derived as

$$\omega_{o,go}^2 = \frac{g_{m3}g_{m4} + g_{oi2}\left(g_{m2} + goi1\right)}{C_1C_2} \approx \omega_o^2 + \frac{g_o g_m}{C^2} \quad (3.25)$$

The simplified results by neglecting very small $go_{i1}go_{i2}$, and the non-zero conductance cause a small frequency shift of the band-pass filter. The quality factor can be expressed as

$$\frac{\omega_{o,go}}{Q_{p,go}} = \frac{g_{m2}}{C_1} + \frac{go_{i1}}{C_1} + \frac{go_{i2}}{C_2} = \frac{\omega_o}{Q_p} + \frac{go_{i1}}{C_1} + \frac{go_{i2}}{C_2} \quad (3.26)$$

From the same high Q condition, we can assume $\omega_{o,go} \approx \omega_o$. Therefore,

$$\frac{1}{Q_{p,go}} = \frac{1}{Q_p} + \frac{1}{\omega_o}\left(\frac{go_{i1}}{C_1} + \frac{go_{i2}}{C_2}\right) = \frac{1}{Q_p} + \frac{2g_o}{\omega_o C} \quad (3.27)$$

Using (3.2), the affected quality factor can be expressed by the relation of the ideal quality factor and the integrator DC gain

$$\frac{1}{Q_{p,go}} = \frac{1}{Q_p} + \frac{2}{A_i} \quad (3.28)$$

Although the finite output conductance will not introduce a large deviation in the center frequency, the gain will be much higher than ideal Q_P to maintain the effective $Q_{p,go}$ performance.

3.2.4.2 Parasitic Poles and Zeros

When the parasitic effect in (3.21) is applied to the integrators, the transfer function in (3.10) becomes

$$H_{nonideal,\tau b}(s) = \frac{s\frac{C_2 g_{m1}(1-s\tau_b)}{C_2(C_1-\tau_b g_{m2})+g_{m3}g_{m4}\tau_b^2}}{s^2 + s\frac{C_2 g_{m2}-2g_{m3}g_{m4}\tau_b}{C_2(C_1-\tau_b g_{m2})+g_{m3}g_{m4}\tau_b^2} + \frac{g_{m3}g_{m4}}{C_2(C_1-\tau_b g_{m2})+g_{m3}g_{m4}\tau_b^2}} \quad (3.29)$$

To simplify the equation, the term τ_b^2 can be neglected since they are very small. This equation also shows a slightly higher low frequency value than the ideal ones. The affected center frequency is therefore given by

$$\omega_{o,\tau b}^2 = \frac{g_{m3}g_{m4}}{C_2\left(C_1 - \tau_b g_{m2}\right)} = \omega_o^2\left(1 + \frac{\omega_o \tau_b}{Q_p}\right) \quad (3.30)$$

Thus, the parasitic effect induces a small shift in ideal center frequency, and it can be neglected for high Q_p implementation. Also, the affected quality factor can be expressed as

$$\frac{\omega_{o,\tau b}}{Q_{p,\tau b}} = \frac{C_2 g_{m2} - 2 g_{m3} g_{m4} \tau_b}{C_2 (C_1 - \tau_b g_{m2}) + g_{m3} g_{m4} \tau_b^2} \approx \frac{g_{m2}}{C_1} + \frac{\tau_b g_{m2}^2}{C_1^2} - \frac{2 g_{m3} g_{m4} \tau_b}{C_1 C_2} = \frac{\omega_o}{Q_p} - \omega_o^2 \tau_b \left(2 - \frac{1}{Q_p^2}\right) \tag{3.31}$$

Equation (3.31) can be simplified under a high Q_p implementation, and the affected quality factor can be calculated as

$$\frac{1}{Q_{p,\tau b}} = \frac{1}{Q_p} - 2\omega_o \tau_b \tag{3.32}$$

The same, parasitic pole and zero would not introduce large deviation of the center frequency, but the equivalent zero should be located at a higher frequency than ideal center frequency, and thus effective $Q_{p,\tau b}$ performance is maintained. It should be noted that the parasitic effects are determined by the finite bandwidth of transconductor g_{m3} and g_{m4}, rather than the small g_{m1} and g_{m2}.

If we take both the finite output conductance and parasitic effects into consideration, the affected ω_o would be very close to the ideal ω_o in high Q_p approximation, and the affected Q_p is given by taking the integrator quality factor in (3.7)

$$\frac{1}{Q_{p,go\&\tau b}} = \frac{1}{Q_p} + \frac{\tau_a}{\omega_o} - 2\omega_o \tau_b = \frac{1}{Q_p} + \frac{2}{Q_{int}(\omega_o)} \tag{3.33}$$

Thus, the affected filter quality factor would be directly related to the integrator non-ideal quality factor.

3.2.4.3 Noise

The integrator output noise is caused by the noise output current of the transconductor, and the circuit design of the transconductor would affect the noise performance. The output noise current power spectral density is given by

$$\overline{i_n^2} = 4kTFg_m \tag{3.34}$$

where k is the Boltzmann constant, T is the absolute temperature and F is a noise factor defined by the transconductor implementation. If F is equal to 1, the noise current is equal to the value generated by an equivalent resistor with the value of $1/g_m$. Figure 3.11a shows the biquad band-pass filter with noisy sources and infinite integrator quality factor. The input terminal is grounded to calculate the noise performance. Since the biquad band-pass filter is the synthesis result from a passive RLC network, the noise output voltage of the filter can be calculated by summing the equivalent output current at the output node of the passive network. From Fig. 3.11b, we can obtain

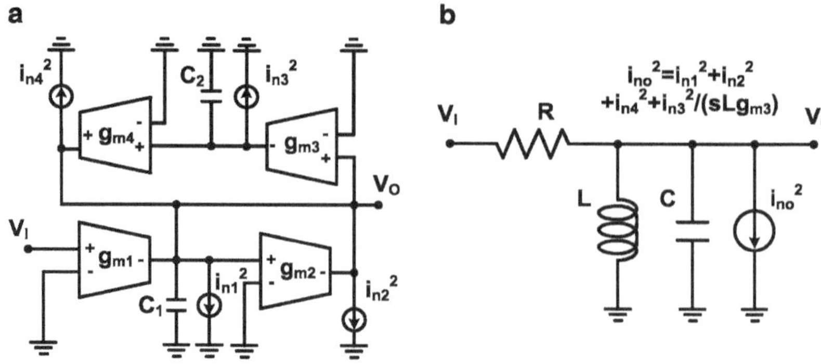

Fig. 3.11 Noise model; (**a**) the biquad bandpass filter including integrator noise model (**b**) the bandpass passive prototype with integrator noise model

$$\frac{\overline{v_{on}^2}(\omega)}{df} = \frac{\left(\overline{i_{n1}^2} + \overline{i_{n2}^2} + \overline{i_{n4}^2}\right) L^2 R^2 \omega^2 + \overline{i_{n3}^2}\left(\frac{R}{g_{m3}}\right)^2}{C^2 L^2 R^2 \omega^4 + L^2 \omega^2 + R^2 - 2LCR^2\omega^2} \tag{3.35}$$

where $L = C_2 / g_{m3} g_{m4}$, $C = C_1$, and $R = 1 / g_{m3} = 1 / g_{m4}$. If the filter is a high Q_p implementation, we use the same assumption and the output noise current generated by g_{m1} and g_{m2} is much smaller than g_{m3} and g_{m4}. This condition yields

$$\frac{\overline{v_{on}^2}(\omega)}{df} = \frac{\left(L^2 R^2 \omega^2 + \frac{LR^2}{C}\right) 4kTFgm}{C^2 L^2 R^2 \omega^4 + L^2 \omega^2 + R^2 - 2LCR^2\omega^2} \tag{3.36}$$

Note that the group delay of the band-pass filter can be expressed as

$$\tau_{group}(\omega) = \frac{-d\varphi(\omega)}{d\omega} = \frac{CL^2 R\omega^2 + LR}{C^2 L^2 R^2 \omega^4 + L^2 \omega^2 + R^2 - 2LCR^2\omega^2} \tag{3.37}$$

where $\varphi(\omega)$ is the phase of the transfer function. Combining (3.36) and (3.37), we have

$$\frac{\overline{v_{on}^2}(\omega)}{df} = \frac{R}{C}\tau_{group}(\omega)(4kTFg_m) = \frac{4kT}{C} Q_p F \tau_{group}(\omega) \tag{3.38}$$

Thus, the filter output noise power spectrum density has the same shape as the group delay of the filter. Since a peak appears at w_o for the group delay, the noise behaves the same way. If a filter transfer function is given, we can reduce the output noise level by increasing C while keeping G_m/C constant or decreasing F from the transconductor circuit design.

The integrated output noise over the frequency range is given by

$$\overline{v_{on}^2}(\omega) = \int_0^\infty \frac{4kT}{C} Q_p F \tau_{group}(\omega)\, df = \frac{4kT}{2\pi C} Q_p F \int_0^\infty \tau_{group}(\omega)\, d\omega = \frac{2kT}{C} Q_p F \tag{3.39}$$

This calculation assumes that F is frequency independent. The noise power is dependent on the quality factor, loading capacitor and noise factor of the transconductor.

For a programmable constant-C_L filter, F would be adjusted to a different programming factor since we change the transconductance and $\tau_{group}(\omega)$ is denormalized to the corresponding value. Thus, the integrated output noise power would still be the same over the frequency tuning range. This effect can also be verified by using the linear time scaling technique [37].

3.2.4.4 Dynamic Range Performance

For analog circuits, some of different performances should meet the system specification. The typical filter performances are the accuracy of the transfer function, power consumption, linearity, and noise. The linearity would limit the largest input signal swing range, and the noise would limit the smallest useful values. Thus, the dynamic range would be determined by both the linearity and noise together.

A. Total Harmonic Distortion (THD)

For linear time-invariant system, the output signal will be linearity related to the input signal. The output signal will have the same frequency component but the magnitude and phase can be different. If the input signal at a specific frequency is applied to non-linear time-invariant system, the output signal would include additional harmonic components. The total harmonic distortion (THD) is defined to be the ratio of the power including second and higher order harmonic terms to the power of the first order component of the signal. It has the following relation

$$THD = 10 \log \left(\frac{V_{h2}^2 + V_{h3}^2 + \cdots}{V_f^2} \right) \tag{3.40}$$

where V_f is the fundamental component and V_{hi} is the amplitude of the ith order harmonic component. The THD can also be presented as a percentage value. Since the THD is given by a ratio, the value should be reported with the combination of an input signal level. In reality, the first five to ten harmonics are calculated owing to that high order terms would be less than the noise power. A THD with the value of −40 dB, which is equivalent to 1%, would be usually required for a G_m-C implementation.

For a non-linear circuit, we apply a sinusoid signal Acos(ω t) to input terminal and then the output can be expanded by a Taylor series formula as

$$\begin{aligned} v_o(t) &= V_f v_{in}(t) + V_2 v_{in}^2(t) + V_3 v_{in}^3(t) + \cdots \\ &= V_f A\cos(\omega t) + V_2 A^2 \cos^2(\omega t) + V_3 A^3 \cos^3(\omega t) + \cdots \end{aligned} \tag{3.41}$$

Since the third-order harmonic term dominates the linearity performance of the differential circuit, we neglect even order distortion components. Equation (3.41) can be approximated as

$$\begin{aligned} v_o(t) &\approx V_f A\cos(\omega t) + V_3 A^3 \cos^3(\omega t) \\ &= \left(V_f A + \frac{3}{4} V_3 A^3\right)\cos(\omega t) + \frac{1}{4} V_3 A^3 \cos(3\omega t) \end{aligned} \tag{3.42}$$

The output signal is composed by a fundamental term and a third-order distortion term (HD3). The HD3 is defined by the ratio of the fundamental term to the third-order distortion term. Usually $(3/4)V_3A^3 \ll V_fA$ can hold and HD3 is given by

$$HD3 = \left(\frac{V_3}{4V_f}\right) A^2 \tag{3.43}$$

To measure the THD or HD3 of a low-pass filter, the input signal frequency should be less than one third of cutoff frequency. If an input signal with higher frequency is applied, the harmonic components would appear at filter stopband, and thus the obtained value is attenuated by the filter. However, the parasitic capacitance will degrade the linearity of transconductor at higher frequency. Then, the filter linearity is also reduced and the filter is worse at highest frequency. The highest frequency, which interests in signal processing of the filter, is the cutoff frequency. To obtain the actual linearity performance of the filter, we should measure the linearity performance at the cutoff frequency and the two-tone test described in the next section is more suitable. The THD is a straight and simple test but it would not work well with signals near the passband.

B. The Third-Order Intercept Point (IP3)

The third-order intercept point (IP3) is a measure for the third-order distortion component. Before we present IP3, the two-tone inter-modulation test should be introduced at first. By using two-tone signals with different frequencies to the input of a circuit, the output signal results from the multiplied input signals. The property can be obtained from a two-tone signal as

$$v_{in}(t) = A_1\cos(\omega_1 t) + A_2\cos(\omega_2 t) \tag{3.44}$$

As the non-linear output is expanded by a Taylor series formula in (3.41), the output signal in this case is given by

$$\begin{aligned}
v_o(t) \approx & V_f\,[A_1\cos(\omega_1 t) + A_2\cos(\omega_2 t)] + V_3\,[A_1\cos(\omega_1 t) + A_2\cos(\omega_2 t)]^3 \\
= & \left(V_f A_1 + \frac{3}{4}V_3 A_1^3 + \frac{3}{2}V_3 A_1 A_2^2\right)\cos(\omega_1 t) \\
+ & \left(V_f A_2 + \frac{3}{4}V_3 A_2^3 + \frac{3}{2}V_3 A_2 A_1^2\right)\cos(\omega_2 t) \\
+ & \frac{3V_3 A_2 A_1^2}{4}\left[\cos(2\omega_1 t + \omega_2 t) + \cos(2\omega_1 t - \omega_2 t)\right] \\
+ & \frac{3V_3 A_1 A_2^2}{4}\left[\cos(2\omega_2 t + \omega_1 t) + \cos(2\omega_2 t - \omega_1 t)\right] \\
+ & \frac{V_3}{4}\left(A_1^3 + A_2^3\right)\left[\cos(3\omega_2 t) + \cos(3\omega_1 t)\right]
\end{aligned} \tag{3.45}$$

In the modulation test, we usually choose $A_1 = A_2 = A$. The two-tone performance is also called the third-order inter-modulation (IM3). The first and second line of (3.45) shows the fundamental component. The third and fourth line of (3.45) shows the distortion at nearly three times the fundamental frequency and two new frequencies that are close to the input frequencies. Thus, the distortion terms can appear at frequencies near the input signal if the frequencies of two-tone signal are very close to each other. The IM3 can be given as

$$A_{IM3} = \frac{A_3}{A_f} = \frac{3V_3A^2}{4V_f} \tag{3.46}$$

where A_f is the amplitude of fundamental component and A_3 is the amplitude of inter-modulation term. The measurement is suitable for narrow band applications such as a band-pass filter because the distortion terms would appear at passband. This test can also be used to obtain the performance of a low-pass filter at cutoff frequency.

Since IM3 is the result of a ratio, we should give the input signal power at the same time to characterize the circuit behavior. On the other hand, the IP3 shows another way. It can be measured when the magnitude of the input signal is small, and thus the higher order distortion terms are negligible. If we neglect the cubic distortion term of the first and second line in (3.45), we can obtain that the magnitude of the fundamental term is proportional to the magnitude of input signal. For the third and fourth lines, the magnitude of inter-modulation terms has a cubic function. Thus, the magnitude of inter-modulation term grows three times than fundamental terms in a logarithmic scale, and the IP3 is defined to be the intersection point of the two lines. Figure 3.12 shows the logarithmic illustration of IP3. The horizontal coordinate of the point is called the input IP3 (IIP3), and it is the input signal

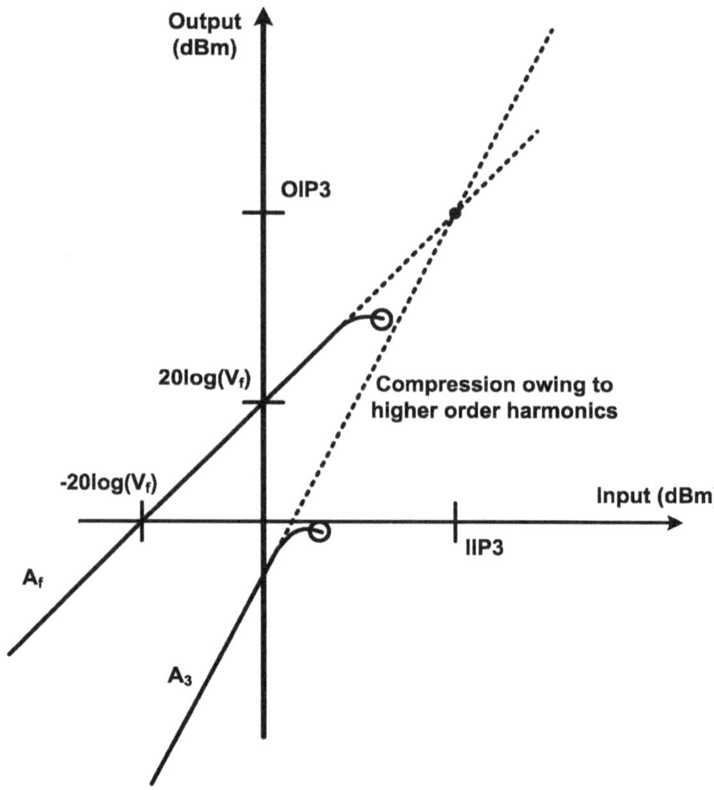

Fig. 3.12 The illustration of input and output intercept point with the unit of decibel

magnitude when the third-order harmonic term is equal to the fundamental terms. The vertical coordinate is called the output IP3 (OIP3). In a low-pass filter design with unity gain transfer function, the IIP3 will be equal to OIP3.

The IP3 not only shows the power of input signal but also the circuit linearity. Recall (3.45), we again assume $A_1 = A_2 = A$, and the IIP3 is given by

$$A_{IIP3} = \sqrt{\frac{4}{3}\left(\frac{V_f}{V_3}\right)} \tag{3.47}$$

Then, the OIP3 is equal to $V_f A_{IP3}$. To obtain the IP3 practically, small input signal is applied to point out the line of third-order distortion component. The reason in that the IP3 is usually out of allowable input range and the value is sometimes larger than the supply voltage in nano-scale CMOS technology. Even though a large input signal can be applied, the higher order non-linearity term would appear and the compression effect occurs.

A useful method to measure IIP3 is discussed. By substituting (3.46) into (3.47), we can obtain $1/A_{IM3} = A_{IIP3}^2/A^2$. In the logarithmic scale, the value can be expressed as

$$20\log(A_{IP3}) = \frac{1}{2}\left[-20\log(A_{IM3})\right] + 20\log(A) \tag{3.48}$$

Therefore, we can denote (3.48) as IIP3 $= A - A_{IM3}/2$ in the unit of decibel (dB), and then OIP3 $= A_f - A_{IM3}/2$. If all the signals are given in dBm, the IP3 is equal to half the absolute value of IM3 plus the corresponding input level. Thus, the IP3 can be measured with only one input level rather than extrapolation.

C. Spurious-Free Dynamic Range (SFDR)

The spurious-free dynamic range is defined as the signal-to-noise ratio when the third-order inter-modulation components are equal to the noise power. This condition happens at the small input signal amplitude. The relationship can be expressed as SFDR $= A_f^* - N_o = A_f^* - A_3^*$, where N_o is the noise power, and A_f^* and A_3^* are the magnitudes of output and inter-modulation component when inter-modulation component is equal to noise power. Using (3.48), we can obtain

$$\text{OIP3} = A_f^* - \frac{\left(N_o - A_f^*\right)}{2} \tag{3.49}$$

Substituting A_f^* = SFDR + N_o in to (3.49), we can have

$$\text{SFDR} = \frac{2\,(OIP3 - N_o)}{3} \tag{3.50}$$

3.3 A Wide Tuning Range G_m-C Continuous-Time Analog Filter

A CMOS operational transconductance amplifier for low-power and wide tuning range filter application is proposed. The transconductor can work from the weak inversion region to the strong inversion region to maximize the transconductance tuning range. The transconductance can be tuned by changing its bias current. A fifth-order Elliptic low-pass filter implemented with the transconductors was integrated by TSMC 180 nm CMOS process. The filter can operate with the cutoff frequency of 250 Hz to 1 MHz. The wide tuning range filter would be suitable for multi-mode applications, especially under the consideration of saving chip areas. The third-order inter-modulation of −40 dB was measured over the tuning range with two tone input signals. The power consumption is 0.8 mW at 1 MHz cutoff frequency and 1.8-V supply with the active area less than 0.3 mm^2.

3.3.1 Introduction

The current trend of the portable solutions tends to include multiple applications in a long-standby system. Cost and power consumption are two most important factors for these products. Cost efficiency has been greatly increased with the emergence of CMOS technology in high performance VLSI implementations. Moreover, to save the silicon area in a multimedia system-on-a-chip solution, re-usable circuits for different system applications can be even cost-effective. For the power consumption, digital circuits can benefit from the supply voltage reduction, but analog circuits can not necessarily decrease the power consumption with the decrease of supply voltage. To meet different specifications for low power consumption, new basic analog building blocks should be re-designed.

In the analog signal processing, the low-pass filter would be one of the most important circuits in the transceiver architecture. There are different ways to implement low-pass filters by CMOS technology at the circuit level. The switched-capacitor (SC) technique which uses switches, capacitors, and operational amplifiers exhibits good linearity, but with the problems of larger power consumption. The active-RC technique which uses operational amplifiers, resistors, and capacitors also exhibits high linearity, but large chip areas will be consumed for resistors or capacitors. MOS varactors were demonstrated to achieve frequency tuning in the active-RC structure [38]. Another technique to realize continuous-time analog filters is to utilize transconductors and capacitors to implement integrators. The absence of the local feedback in the G_m-C analog filter technique performs good frequency responses of the signal transfer functions [39]. Furthermore, G_m-C analog filters do not require extra processing steps, as compared with active-RC filters, and their frequency tuning is easily achieved using DC bias voltage or current. The performance of the transconductor will largely affect the G_m-C analog filters. Many of the previously published papers made efforts on improving the speed, linearity, or dynamic range of transconductor circuits.

This work presents a CMOS implementation of a low-power fifth-order Elliptic low-pass G_m-C filter for a very wide frequency tuning range, which can operate as the channel selection filter for the audio, speech, bio-medical, and wireless application. The high performance V-I circuit which can operate in both weak inversion and strong inversion regions is designed to achieve a wide transconductance tuning range. The working mode of the transconductor can be set in different inversion regions and the transconductance can be widely tuned by changing DC bias current. The transconductor is used as an operational transconductance amplifier in the design of the fifth-order Elliptic low-pass G_m-C filter. In the section, Section 3.3.2 develops the low power and wide tuning range transconductor architecture. The equivalent resistor used in the proposed transconductor is discussed in Section 3.3.3. The G_m-C filter is developed in Section 3.3.4, followed by the experimental results of both the transconductor and the filter in Section 3.3.5. Summary is presented in Section 3.3.6.

3.3.2 The Proposed Transconductor Cell

3.3.2.1 Implementation of Linearization Technique

In order to increase the tuning range of the filter, the range of the transconductance should be increased first. The proposed transconductor circuit based on the translinear loop is shown in Fig. 3.13. The MOSFETs M1 to M9 should be appropriately designed to operate in both weak and strong inversion regions in order to increase the transconductance range. The linearity has been well maintained while tuning the transconductance.

3.3.2.2 The Transconductor Cell Operating in the Weak Inversion Region

For a MOSFET operating in the weak inversion region with V_{DS} larger than a few times of thermal voltage U_T, its current exhibits an exponential dependence of V_{GS}, as shown in (3.51),

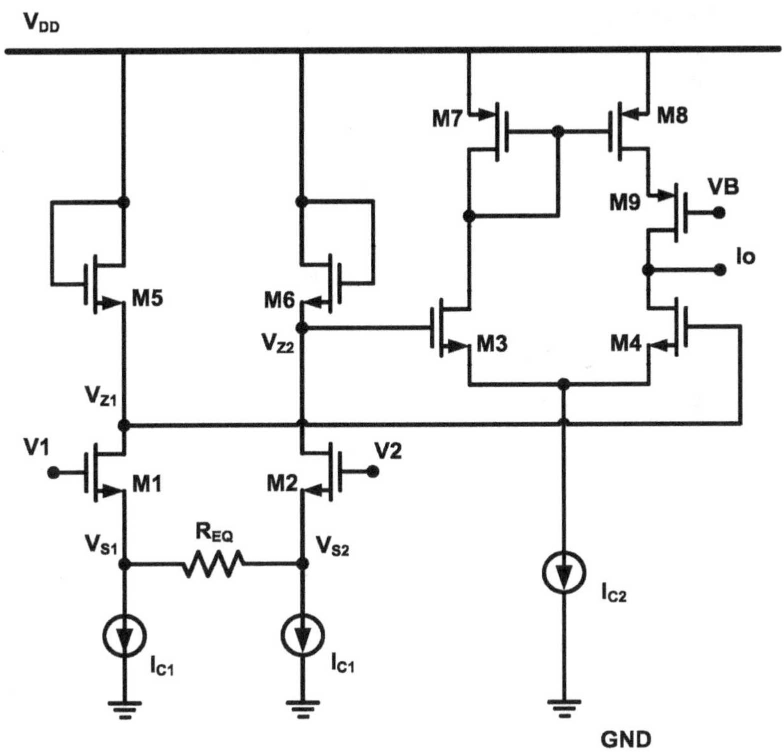

Fig. 3.13 The proposed transconductor circuit

$$I_D = I_{D0}\frac{W}{L}\exp\left(\frac{V_{GS}}{nU_T}\right) \tag{3.51}$$

where W and L are the width and the length of the transistor, respectively, I_{D0} is the reverse saturation current, n is the subthreshold slope factor, and U_T is the thermal voltage. With the weak inversion characteristic, the input voltage should be logarithmically determined first and through the exponential function, the output current can then be linearized, as shown in transistors M1 to M6 of Fig. 3.13. When both the input and the output stages of the transconductor operate in the weak inversion mode (Weak-IN Weak-OUT mode), and the same transistor sizes for M1 to M6 are used, the relationship between differential input voltage and output current can be derived and analyzed. The diode-connected transistors M5 and M6 carry the current I_{D5} and I_{D6}, respectively, and we assume the two gate voltage of transistors M3 and M4 are V_{z2} and V_{z1}. $V_{z1} - V_{z2}$ can be obtained by

$$V_{z1} - V_{z2} = (V_{DD} - V_{gs5}) - (V_{DD} - V_{gs6}) = V_{gs6} - V_{gs5} \tag{3.52}$$

$$V_{z1} - V_{z2} = nU_T ln\left(\frac{I_{D6}}{I_{D5}}\right) \tag{3.53}$$

For current I_{D3} and I_{D4}

$$\frac{I_{D4}}{I_{D3}} = exp\left(\frac{V_{z1} - V_{z2}}{nU_T}\right) \tag{3.54}$$

From Equations (3.53) and (3.54), we can obtain

$$I_o = I_{D3} - I_{D4} = \frac{I_{C2}}{I_{C1}}\frac{(I_{D5} - I_{D6})}{2} \tag{3.55}$$

To make sure a linear voltage to current conversion is achieved, a resistor R_{eq} connected at the source terminals of transistors M1 and M2 is introduced. The resistor should be chosen much larger than U_T/I_{C1} so that we can have

$$\frac{(I_{D5} - I_{D6})}{2}R_{eq} = V_1 - V_2 \tag{3.56}$$

By substituting (3.56) into (3.55), we can obtain

$$I_o = \frac{I_{C2}}{I_{C1}}\frac{(V_1 - V_2)}{R_{eq}} \tag{3.57}$$

The output current can be tuned by two DC bias current I_{C1} and I_{C2} for various transconductance values. In the weak inversion structure, the output current will be small due to the large resistor R_{eq} and thus it will be strongly affected by the output DC voltage. Transistor M9 is added to reduce the channel length modulation effect from the current mirror M7 and M8 because of the different V_{DS} voltage.

In order to take nonlinearity performance into consideration, from a Taylor series analysis, it turns out that the total harmonic distortion of the proposed circuit is dominated by the third-order harmonic distortion measure of this V-I conversion. The HD3 could be approximated as

$$HD_3 = \frac{2\left(I_{C1}R_{eq}\right)^2 nU_T}{\left(2I_{C1}R_{eq} + nU_T\right)^4\left(2I_{C1}R_{eq} + 2nU_T\right)}(V_1 - V_2)^2 \tag{3.58}$$

As the equation shown, the linearity performance can be improved by increasing the tail current I_{C1} and the equivalent resistor R_{eq}. However, the increased I_{C1} would lead to higher power consumption, and thus we should choose smaller I_{C1} and larger R_{eq} in order to achieve low-power and high linearity. As long as the resistance of R_{eq} can be kept large enough, in the order of Mega ohm, a linear transconductor operation in the weak inversion region can be accomplished. In addition, it is noted that I_{C2} does not affect the linearity performance in the circuit.

3.3.2.3 The Transconductor Cell Operating in the Strong Inversion Region

As the transistors M1 to M6 are working in the strong inversion region (Strong-IN Strong-OUT mode) with the same sizes, the relationship between the output current and the input voltage can be derived as follows.

Again, $V_{z1} - V_{z2} = V_{gs6} - V_{gs5}$. Because the drain current of M1 and M5 are the same and the drain current of M2 and M6 are the same,

$$(V_{gs5} - V_T)^2 = (V_{gs1} - V_T)^2 = (V_1 - V_{s1} - V_T)^2 \tag{3.59}$$

$$(V_{gs6} - V_T)^2 = (V_{gs2} - V_T)^2 = (V_2 - V_{s2} - V_T)^2 \tag{3.60}$$

$$V_{gs5} = V_{gs1} = V_1 - V_{s1} \tag{3.61}$$

$$V_{gs6} = V_{gs2} = V_2 - V_{s2} \tag{3.62}$$

From (3.59) to (3.62), assume $I_{D5} > I_{D6}$ and the current flowing through R_{eq} is ΔI,

$$\begin{aligned} V_{z1} - V_{z2} &= (V_2 - V_{s2}) - (V_1 - V_{s1}) \\ &= (V_2 - V_1) - (V_{s2} - V_{s1}) \\ &= (V_2 - V_1) + R_{eq}\Delta I \end{aligned} \tag{3.63}$$

$$V_1 - V_2 = (V_{z2} - V_{z1}) + R_{eq}\Delta I \tag{3.64}$$

For the current through transistors M5 and M6, we can have

$$I_{D5} = I_{C1} + \Delta I = \frac{K}{2}(V_{gs1} - V_T)^2 \tag{3.65}$$

$$I_{D6} = I_{C1} - \Delta I = \frac{K}{2}(V_{gs2} - V_T)^2 \tag{3.66}$$

where $K = \mu_n C_{ox}(W/L)$, W and L are the width and length of the device, respectively, C_{ox} is the oxide capacitance per unit channel area, and μ_n is the low-field mobility. Therefore,

$$V_{gs1} = \sqrt{\frac{2(I_{C1} + \Delta I)}{K}} + V_T \tag{3.67}$$

$$V_{gs2} = \sqrt{\frac{2(I_{C1} - \Delta I)}{K}} + V_T \tag{3.68}$$

From (3.61), and (3.62), we can find

$$\begin{aligned} V_{z2} - V_{z1} &= V_{gs1} - V_{gs2} \\ &= \sqrt{\frac{2}{K}}(\sqrt{I_{C1} + \Delta I} - \sqrt{I_{C1} - \Delta I}) \\ &= \sqrt{\frac{2I_{C1}}{K}}\left(\sqrt{1 + \frac{\Delta I}{I_{C1}}} - \sqrt{1 - \frac{\Delta I}{I_{C1}}}\right) \end{aligned} \tag{3.69}$$

If we assume $\Delta I/I_{C1} \ll 1$, i.e., $\Delta I \ll I_{C1}$,

$$\begin{aligned} V_{z2} - V_{z1} &= \sqrt{\frac{2I_{C1}}{K}}\left[\left(1 + \frac{\Delta I}{2I_{C1}}\right) - (1 - \frac{\Delta I}{2I_{C1}})\right] \\ &= \sqrt{\frac{2}{KI_{C1}}}\Delta I \end{aligned} \tag{3.70}$$

Substituting (3.70) into (3.64), we obtain

$$V_1 - V_2 = \sqrt{\frac{2}{KI_{C1}}}\Delta I + R_{eq}\Delta I \tag{3.71}$$

Therefore,

$$\Delta I = \frac{1}{\sqrt{\frac{2}{KI_{C1}}} + R_{eq}}(V_1 - V_2) \tag{3.72}$$

$$V_{z1} - V_{z2} = \frac{\sqrt{\frac{2}{KI_{C1}}}}{\sqrt{\frac{2}{KI_{C1}}} + R_{eq}}(V_1 - V_2) \tag{3.73}$$

The value of the equivalent resistor R_{eq} should be chosen much smaller than the equivalent resistor in the weak inversion structure. If the equivalent resistor R_{eq} were too large, for example, in the order of Mega ohm, Δ I would be almost zero. The voltage difference between V_{z1} and V_{z2} would be zero as well, so there would be no current flowing to the output. The value of the equivalent resistor R_{eq} should be made comparable to the other term in the denominator of (3.73), so the difference $\Delta V_z = (V_{z1} - V_{z2})$, would be large enough to produce output current. From the equations of the differential amplifier the output current I_o can be obtained by

$$I_o = I_{D3} - I_{D4} = \frac{K}{2}(\Delta V_z)\sqrt{\frac{4I_{C2}}{K} - \Delta V_z^2} \tag{3.74}$$

For a small ΔV_z^2 compared to $4\ I_{C2}/K$,

$$I_o \approx \frac{K}{2}(\Delta Vz)\sqrt{\frac{4I_{C2}}{K}} = \frac{\sqrt{\frac{2I_{C2}}{I_{C1}}}}{\sqrt{\frac{2}{KI_{C1}}} + R_{eq}}(V_1 - V_2) \tag{3.75}$$

The relationship between the output current and the input differential voltage can be tuned by I_{C1} and I_{C2}. With a suitable value of I_{C1}, we can increase I_{C2} to increase the overall transconductance. In addition, in the strong inversion region, the HD3 of the proposed transconducor can also be analyzed by a Taylor series expansion and approximated as

$$HD_3 = \frac{1}{4\left(\mathrm{R}_{eq} + \sqrt{\frac{2}{KI_{C1}}}\right)^2 I_{C1}I_{C2}}(V_1 - V_2)^2 \tag{3.76}$$

Thus, the low power and the high linearity performance call for increasing R_{eq}.

3.3.2.4 The Transconductor Cell Operating in the Multi-Inversion Regions

In order to increase the tuning range of the transconductance, the transconductor is designed to be able to switch from the weak inversion region to the strong inversion region continuously. This means that the largest transconductance in the weak inversion region should be larger than the smallest transconductance in the strong inversion region. (The detailed analyses of bias current conditions will be shown in Section III.) Thus, the transconductor circuit would operate in the regions where the input stage, including transistors M1, M2, M5, and M6, remains in the same inversion condition, weak or strong, while the output stage, including transistors M3, M4, M7, M8 and M9, is forced to move from weak to strong inversion region or vice versa.

If we need to increase the transconductance when the transconductor works in the weak inversion region, we should increase the value of I_{C2} to move from Weak-IN Weak-OUT to Weak-IN Strong-OUT operation. At certain point, the increased value of I_{C2} will enable M3, M4, M7, M8 and M9 to enter the strong inversion region. The relationship between output current and the input differential voltage can be approximated to

$$I_o = \frac{\sqrt{KI_{C2}}}{I_{C1}} nU_T \frac{2}{R_{eq}}(V_1 - V_2) \tag{3.77}$$

On the other hand, if we need to decrease the transconductance when the transconductor works in the strong inversion region, we should decrease the value of I_{C2}. At certain point, the decreased value of I_{C2} will have M3, M4, M7, M8 and M9 to enter the weak inversion region. The relationship between the output current and the input differential voltage can now be approximated to

$$I_o = \frac{I_{C2}}{nU_T} \frac{\sqrt{\frac{2}{KI_{C1}}}}{2\left(\sqrt{\frac{2}{KI_{C1}}} + R_{eq}\right)}(V_1 - V_2) \tag{3.78}$$

However, the nonlinear term in the voltage to current relationship will increase in the transition of multi-mode inversion. If we need to achieve a large tuning range and an acceptable linearity, the input voltage swing should be kept limited when operation in multi-mode inversion.

3.3.2.5 Noise Analysis of the Proposed Transconductor

Flicker noise is the dominant input-referred noise source for low frequencies. Using the Flicker noise model of transistors reported in [40], the low frequency noise referred to the input terminals of transconductors can be calculated and its equivalent noise power spectral density (PSD) is given by

$$\begin{aligned} \frac{V_{f-in}^2}{\Delta f} &= 2v_{M1}^2 + v_{\mathrm{R}_{eq,f}}^2 + 2v_{tail}^2 g_{mtail}^2 \mathrm{R}_{eq}^2 + \\ &\quad 2\left[v_{M5}^2 + v_{M3}^2 + v_{M7}^2\left(\frac{g_{m7}}{g_{m3}}\right)^2\right]\left[\frac{\left(1 + g_{m1}\mathrm{R}_{eq}\right) g_{m5}}{g_{m1}}\right]^2 \end{aligned} \tag{3.79}$$

where g_{mi} is the small signal transconductance of transistor M_i, g_{mtail} is the transconductance of the MOS current sources, $V_{Req,f}^2$ is the Flicker noise contributed by the equivalent resistor, V_{tail}^2 is the Flicker noise contributed by the tail current source I_{C1}, and $V_{Mi}^2 = K_F I_D / W_i L_i f$, K_F being the Flicker noise coefficient, f being frequency, W_i and L_i being the width and length of transistor M_i, is the Flicker noise contributed by transistor M_i. Thus, low frequency noise can be minimized by proper design of transistor dimensions and bias current source.

As frequency higher than noise corner frequency, estimated around 300 Hz, the noise source is dominated by the thermal noise. By the thermal noise model given in [25], the input-referred thermal noise can be expressed by

$$\frac{V_{th\text{-}in}^2}{\Delta f} = 8kT\gamma_a\left(\frac{1}{g_{m1}} + g_{m,tail}\mathrm{R}_{eq}^2 + g_{m5}\left[\frac{\left(1+g_{m1}\mathrm{R}_{eq}\right)}{g_{m1}}\right]^2\right) + 8kT\gamma_b\left(\frac{1}{g_{m3}} + \frac{g_{m7}}{g_{m3}^2}\right)\left[\frac{\left(1+g_{m1}\mathrm{R}_{eq}\right)g_{m5}}{g_{m1}}\right]^2 + I_{\mathrm{R}_{eq,th}}^2\mathrm{R}_{eq}^2 \tag{3.80}$$

where k is the Boltzmann constant, T is the temperature, γi is the noise parameter depending on the device bias condition, and $I_{\mathrm{Req,th}}^2$ is the thermal noise contributed by the equivalent resistor. According to [25], the noise parameter γ would be equal to 1/2 and 2/3 at weak and strong inversion regions, respectively. From the equation shown above, a high value of R_{eq}, which improves the linearity, would also induce increased noise. Thus, large aspect ratios of input transistors and small aspect ratios of load and tail current transistors should be chosen for smaller thermal noise power spectral density.

3.3.3 *The Equivalent Resistor R_{EQ}*

3.3.3.1 Switching Methodology

According to the above analysis of different requirement of the equivalent resistor, we would use two different scales of resistors for linearity consideration. Figure 3.2 shows the equivalent circuit of resistor R_{eq}. The resistor can be modeled by the parallel connection of a large resistor and a small resistor. For the constraint of limited chip areas, to implement the large resistor, the equivalent resistor circuit is implemented by CMOS, instead of poly resistors, and we use the output impedance r_o of saturated transistors in our design to emulate the large resistance. The nodes V_{S1} and V_{S2} are connected to the source terminals of M1 and M2 in Fig. 3.13. When M1, M2, M5 and M6 operate in the weak inversion region, V_{mode} is set low to turn off MRS and the equivalent resistance is the output impedance of four parallel connected MOSFETs in the saturation region so as to provide the high resistance. In Fig. 3.14, the transistors MR1 to MR8 working in the saturation region will be used as the equivalent resistor when the transconductor operates in the weak inversion region. V_{S1} and V_{S2} are set to half of the supply voltage to achieve largest swing ranges. Under the DC voltage constraint, to maintain the correct operation region of the equivalent saturated resistor circuit, MS1 to MS4 in Fig. 3.15 are introduced as the level shifter circuit. Thus, the output common mode voltage of the saturated

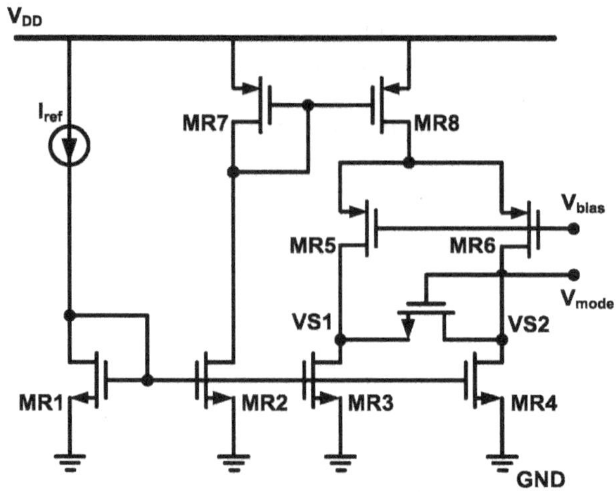

Fig. 3.14 Equivalent resistor circuit

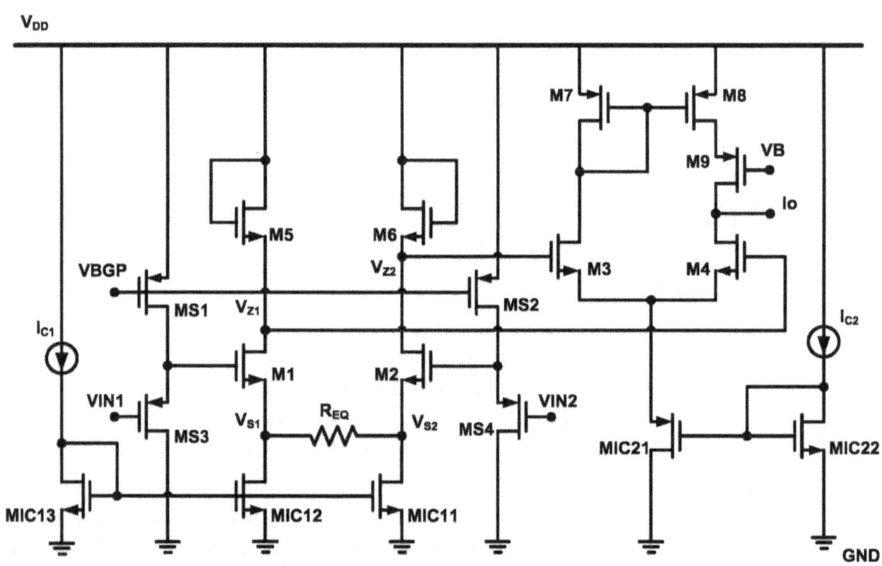

Fig. 3.15 Final implementation of the proposed transconductor circuit

resistor would be approximately equal to the input common mode voltage of the transconductor cell for the largest swing range operation. On the other hand, when M1, M2, M5, and M6 work in the strong inversion region and V_{mode} is set high, a low-value resistor is obtained by the transistor MRS working in the triode region.

3.3.3.2 Bias Current Condition

In order to make sure that the transconductor can continuously operate over the entire range, we should properly design the values of the linear and the saturated resistors.

For the saturated resistors, the conductance can be approximated by [41]

$$\frac{1}{R_{eq,\ large}} = g_o = \frac{I_{ref} X_D \sqrt{K_a}}{4L\sqrt{V_X}} + \sigma\sqrt{K_R I_{ref}} \tag{3.81}$$

where σ is inversely proportional to the cube of the transistor length, Ka is the kappa voltage, X_D is a device fitting parameter, and K_R is the device parameter of MR5 and MR6. The voltage V_X is given by

$$V_X = V_{DS} - V_{DSAT} + \frac{1}{K_a}\left(\frac{E_a X_D}{2}\right)^2 \tag{3.82}$$

where V_{DS} is the drain-source bias voltage and Ea is the Early voltage. From the equation shown above, the resistance can be tuned by the bias current I_{ref}. Moreover, in the sub-micron processes with small device length, the second term in (3.81) dominates the saturated conductance.

The conductance mismatch between bias transistors MR5 and MR6 only slightly affects the saturated resistor performance. The mismatch problem can be solved by choosing larger feature sizes of transistors and careful layout. For parasitic capacitance, analyses indicate that the effect caused by mismatch of gate-drain and drain-source capacitors between MR5 and MR6 can be reduced by a larger capacitor connected to the Vbias node. And for stability analyses, the gate-drain and drain-bulk capacitors will only induce a high frequency pole for the structure. It has been shown by simulation that the pole appears at the GHz range, and thus the saturated resistor is suitable for our frequency range design.

As the MOS transistor is biased in the linear region, the conductance can be expressed as

$$\frac{1}{R_{\mathrm{eq,small}}} = K_r(V_{\mathrm{mode}} - V_{\mathrm{T}} - V_{\mathrm{cms}}) \tag{3.83}$$

where V_T is the threshold voltage, K_r is the device parameter of MRS, and $V_{cms} = (VS1 + VS2)/2$.

As it was mentioned in the previous section, in order to increase the tuning range of the transconductance, the largest transconductance in the weak inversion region should be larger than the smallest transconductance in the strong inversion region. That means that the output current in (3.77) should be larger than the current of (3.78). We can substitute the conductance obtained in (3.81) and (3.83) into (3.77) and (3.78) respectively, and we can find the following limitation

$$\frac{\beta}{\alpha} < \frac{\sqrt{2K}\,(nU_T)^2}{\sigma\sqrt{K_R I_{ref}}} \tag{3.84}$$

where

$$\alpha = \frac{I_{C2\_strong}}{I^2_{C1\_weak}} \tag{3.85}$$

$$\beta = \frac{\sqrt{2} + \frac{1}{K_r(V_{\text{mode}} - V_{\text{T}} - V_{\text{cms}})\sqrt{KI_{C2\_weak}}}}{I_{C1\_strong}} \tag{3.86}$$

α is defined when the input stage stays in the weak inversion region and the output stage enters the strong inversion region, as shown in (3.85), and β is also defined when the input stage stays in the strong inversion region and the output stage enters the weak inversion region, as shown in (3.86).

In addition, the working mode of the circuit can be selected by the gate voltage of MIC13 in Fig. 3.16. If the gate voltage of MIC13 is larger than the threshold voltage, V_T, V_{mode} will be set by series of inverters to supply voltage. Otherwise, if the gate voltage of MIC13 is smaller than V_T, V_{mode} will be set to ground voltage. It is known that the larger transconductance can induce higher cutoff frequency in analog low-pass filter design. In our circuit, the transconductance in both inversion regions will be changed by I_{C1} and I_{C2}.

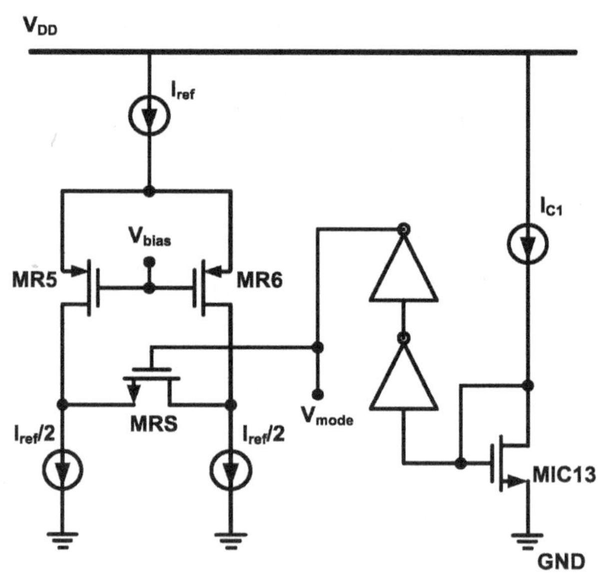

Fig. 3.16 V_{mode} switching circuit for the proposed transconductor

3.3.3.3 Linearity and Noise Analyses

Based on the non-ideal effect of the active device compared with the passive device, the linearity performance of active resistors should be taken into consideration. For the saturated resistor structure, the even-order distortions are cancelled by the differential structure. Thus, third-order harmonic distortion dominates the linearity performance and can be further computed as [41]

$$HD3\_R_{eq,large} = \frac{1}{16}\left(\frac{g_{oL}}{g_{oL} + \sigma\sqrt{K_R I_{ref}}}\right)\left(\frac{V_{S1} - V_{S2}}{V_X}\right)^2 \quad (3.87)$$

where g_{oL} is the conductance expressed in the first term of (3.81). Therefore, the linearity performance can be improved by using larger feature size of the transistors MR5 and MR6.

On the other hand, the even-order distortion of the linear region MOS transistor based resistor would be canceled out. The third-order distortion can be demonstrated by taking body effects into consideration

$$\begin{aligned} HD3\_R_{eq,small} = & \frac{1}{(V_{\text{mode}} - V_{\text{T}} - V_{\text{cms}})\,(2\varphi_F - V_B + V_{\text{cms}})^{\frac{3}{2}}} \\ & \times \frac{\gamma_b}{384}\,(V_{S1} - V_{S2})^2 \end{aligned} \quad (3.88)$$

where γ_b is the body factor, φ_F is the Fermi level, and V_B is the bulk voltage. From (3.88), higher gate voltage or lower bulk voltage of the MOS resistor would result in better linearity. Figure 3.17 shows the harmonic distortion for the linear region and the saturated MOS resistors. The dimensions of transistors MR5 and MR6 are 8 µm/4µm. For MRS, the dimension is 1 µm/1 µm. A 2% mismatch for MR5 and MR6 has been included in the simulation. The distortion level below −40 dB of both resistors can be achieved for a 10 MHz 0.8 V_{pp} input signal. Besides, the short channel effect, due to the electrical and lateral electric fields, also limits the linearity performance. The noise produced by the equivalent resistor circuit will directly affect the noise performance from the analysis of (3.79) and (3.80). For the saturated resistor, the noise is contributed by the addition of transistors MR3, MR4, MR5, and MR6. Transistor MRS induces the linear region MOS resistor noise. Thus, the Flicker and thermal noise power spectral density produced by both the large and small resistors can be given by

$$v^2_{\text{R}_{eq,f}} = \left(2g^2_{mR3}v^2_{MR3} + 2v^2_{MR5}g^2_{mR5}\right)\text{R}^2_{eq} \quad (3.89)$$

$$I^2_{\text{R}_{eq,th}} = 4kT\,[2\gamma_c\,(g_{mR3} + g_{mR5}) + \gamma_d\,g_{mRS}] \quad (3.90)$$

Thus, the optimized values versus the noise and linearity performance would be chosen and adopted carefully from their performance tradeoff.

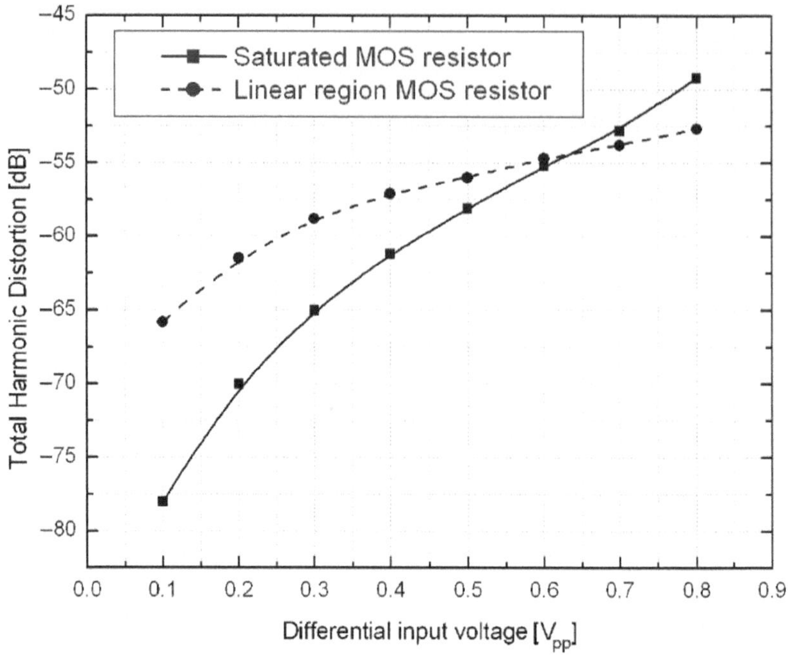

Fig. 3.17 Simulated linearity performance of the equivalent resistor

3.3.4 Filter Architecture

To demonstrate the basic building block in a system level, a fifthorder Elliptic low-pass filter is implemented [42]. The fifth-order Elliptic low-pass filter design starts from a standard fifth-order Elliptic low-pass LC-ladder prototype, as shown in Fig. 3.18. A −6 dB DC gain can be easily obtained under very low frequency when resistor R_S equals to resistor R_L. From the RLC-ladder prototype, through the use of the signal-flow graph method [43], the fifth-order Elliptic low-pass G_m-C filter, which consists of seven identical transconductors and seven capacitors, including two floating capacitors, is obtained. The transconductance, which equals to the value of $1/R_S$ and $1/R_L$, is used for all transconductors. Q tuning circuits are not considered here with the intrinsic quality of the low Q structure. The final G_m-C filter implementation is shown in Fig. 3.19. The transconductor introduced in the previous section is used here in the design of the fifth-order Elliptic low-pass G_m-C filter.

In low-pass filter design, the cutoff frequency of the filter is proportional to g_m/C, where g_m is the transconductance of the transconductor and C is the capacitance. Low cutoff frequency filters are very important in the speech signal processing and medical hearing application. However, there are some problems encountered in these low frequency filter design. The main issues are the large time constant involved and the values of the resistors and capacitance limited by the silicon area.

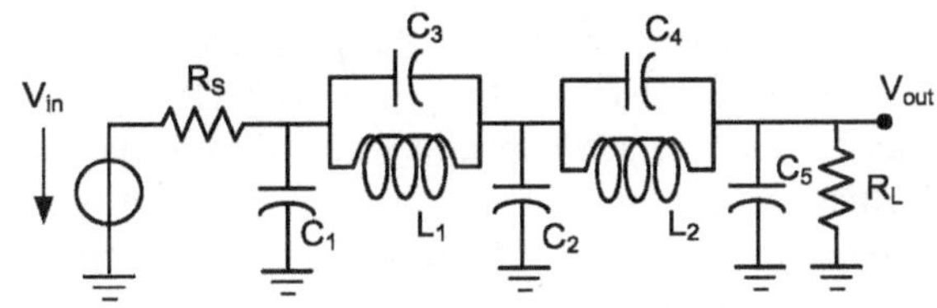

Fig. 3.18 Fifth-order elliptic RLC ladder network prototype

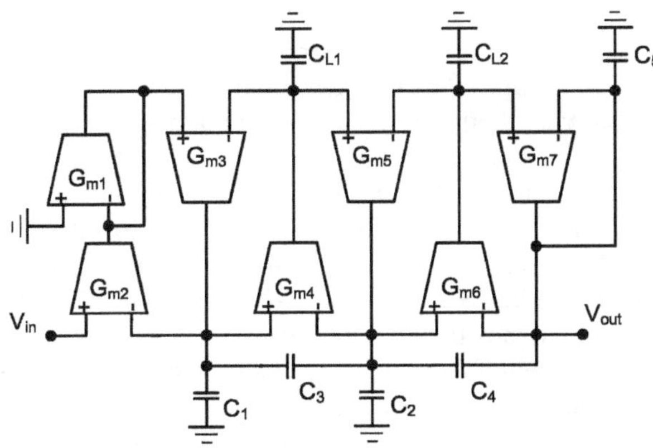

Fig. 3.19 Fifth-order Elliptic low-pass G_m-C filter

In our circuit, when the transconductor works in the weak inversion region, a small value of capacitance will be enough to achieve the low cutoff frequency owing to the small transconductance in the nS order. On the other hand, the larger transconductance can also be obtained in the same filter architecture for higher cutoff frequency when the transconductor works in the saturation region.

The cutoff frequency of the G_m-C filter is tuned by changing the DC bias current of the transconductor. Because the transconductance range of the transconductor is quite wide, the resultant filter also has a wide cutoff frequency range under suitable working mode selection. Tuning circuitry can be developed with digitally controlled circuits in a system-on-a-chip solution by choosing a number of current sources for G_m tuning. The maximum capacitance shown in Fig. 3.19 is only 3.6 pF, so it is easy to integrate with the other circuits.

3.3.5 Experimental Results

The transconductor and the filter were fabricated in the TSMC 180 nm Deep N-WELL CMOS process. Body effects can be simply eliminated by connecting the source and the bulk terminals together in the process. The aspect ratio of 11.5 μm/2 μm is used for transistors M1, M2, M3, M4, M5, and M6, and 11.5 μm/0.2 μm is used for transistors M7, M8, and M9. The value of 8 μm / 4 μm is used for MR5 and MR6, and I_{ref} is equal to 20 μA in the equivalent resistor circuit. 1 μm / 1 μm is used for linear region MOS resistor MRS. Since the mismatches in transconductors and capacitors directly translate to the degradation of overall performance, such as nonlinear effects and errors of transformation, the filter has been laid out very carefully. Metal-insulator-Metal capacitors were used in the circuit and the unit of the capacitor array is 0.1 pF. In this section, the experimental results are presented. All of the results were obtained with a single power supply voltage of 1.8-V.

The measurement results of the transconductor's transfer curves are shown in Fig. 3.20a, b, c and d. In Fig. 3.20a, the measurements of voltage to current transfer curves are obtained when both the input and output stages of the transconductor operate in the weak inversion region as I_{C1} equals to 10 nA and I_{C2} changes from 10 nA to 80 nA. The saturated resistor circuit is selected here by setting V_{mode} to GND. The transfer curves in the weak inversion region follow the expected formula described previously. Also, the transconductance is changed from 2.5 nS to 16 nS, as shown in the figure. In Fig. 3.20b, owing to the increment of I_{C2}, the output stage of the transconductor extends the operation from the weak inversion region to the strong inversion region while the input stage and the equivalent resistor remain in the same previous condition, and thus achieves to a much larger transconductance. The measured transconductance could be tuned from 3 nS to 610 nS by increasing I_{C2} to the value of 10 μA.

In Fig. 3.20c, the input and output stages of the transconductor both operate in the strong inversion region to achieve a large transconductance, as compared with the weak inversion region operation. The linear region MOS resistor circuit is selected here by setting V_{mode} to VDD. The transconductance from 11 μS to 18 μS is obtained while the current I_{C1} equals to 1 μA and I_{C2} changes from 10 μA to 40 μA. We can find that the transconductance would be tuned proportional to the square root of I_{C2} as described in the formula above. In Fig. 3.20d, the decreased I_{C2} results that the output stage of the transconductor extends the operation from the strong inversion region to the weak inversion region while the input stage and the equivalent resistor remain in the same condition, and thus achieves to a much smaller transconductance. As shown in the figure, the transconductance could be tuned from 20.2 μS to 0.19 μS by decreasing I_{C2} to the value of 0.5 μA.

From the measurement results of Fig. 3.20b and 3.20d, continuous tuning from weak inversion operation to strong inversion operation can be guaranteed and the transconductor can be tuned for a very wide range. However, the linearity of the transconductor should be maintained over the range because it will directly affect the linearity of the proposed G_m-C filter. For the linearity of this transconductor, when both the input and output stages of the transconductor operate in the weak inversion region, THD is about −56 dB with $I_{C1} = 10$ nA $I_{C2} = 10$ nA at 100 Hz 300 mV_{pp} input signal. As I_{C2} increases to 10 μA, the output stage will operate in the strong inversion region while the input stage stays in the weak inversion region, and THD is measured to be −44 dB. On the other hand, when the input and output stages of the transconductor operate in the strong inversion region, THD of −43 dB is measured by giving $I_{C1} = 1$ μA and $I_{C2} = 40$ μA with 10 KHz 300 mV_{pp} input

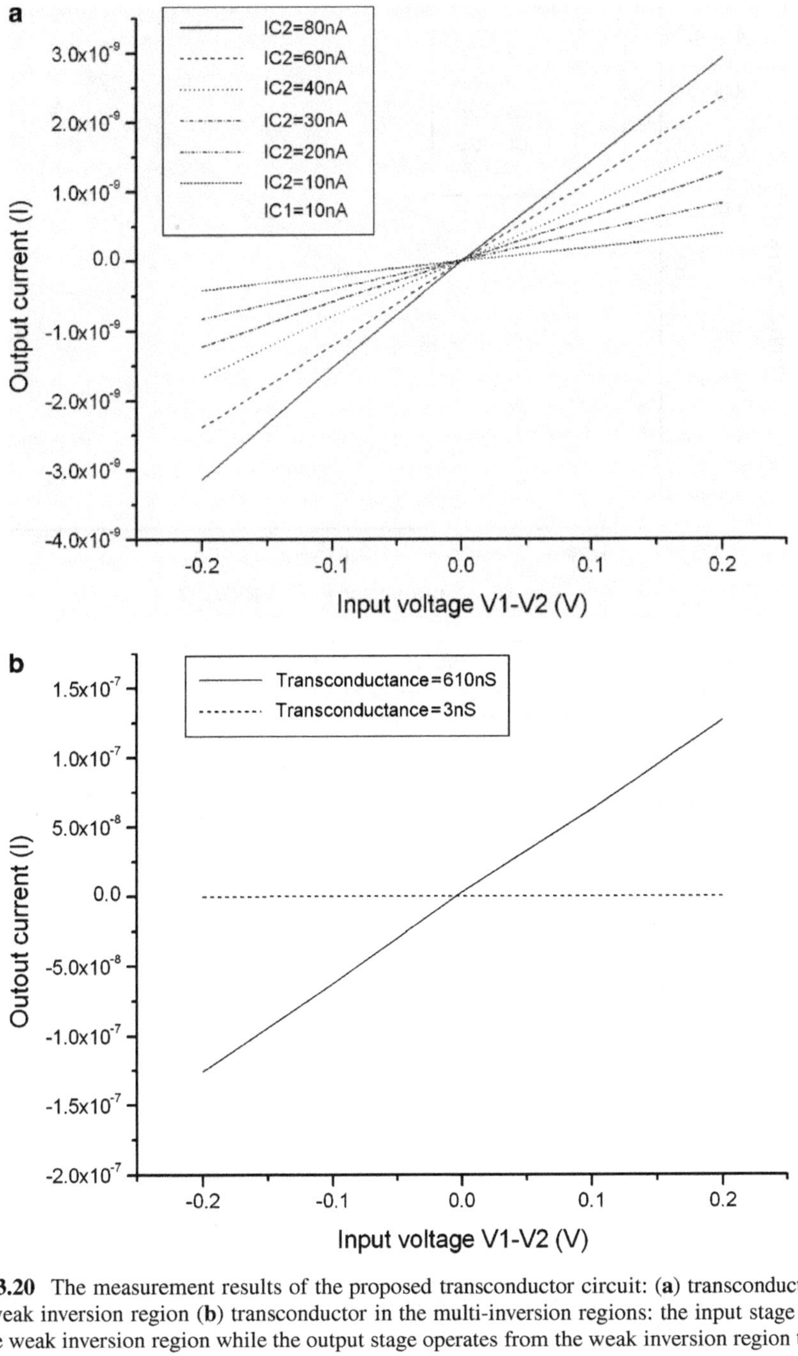

Fig. 3.20 The measurement results of the proposed transconductor circuit: (**a**) transconductor in the weak inversion region (**b**) transconductor in the multi-inversion regions: the input stage stays in the weak inversion region while the output stage operates from the weak inversion region to the strong inversion region

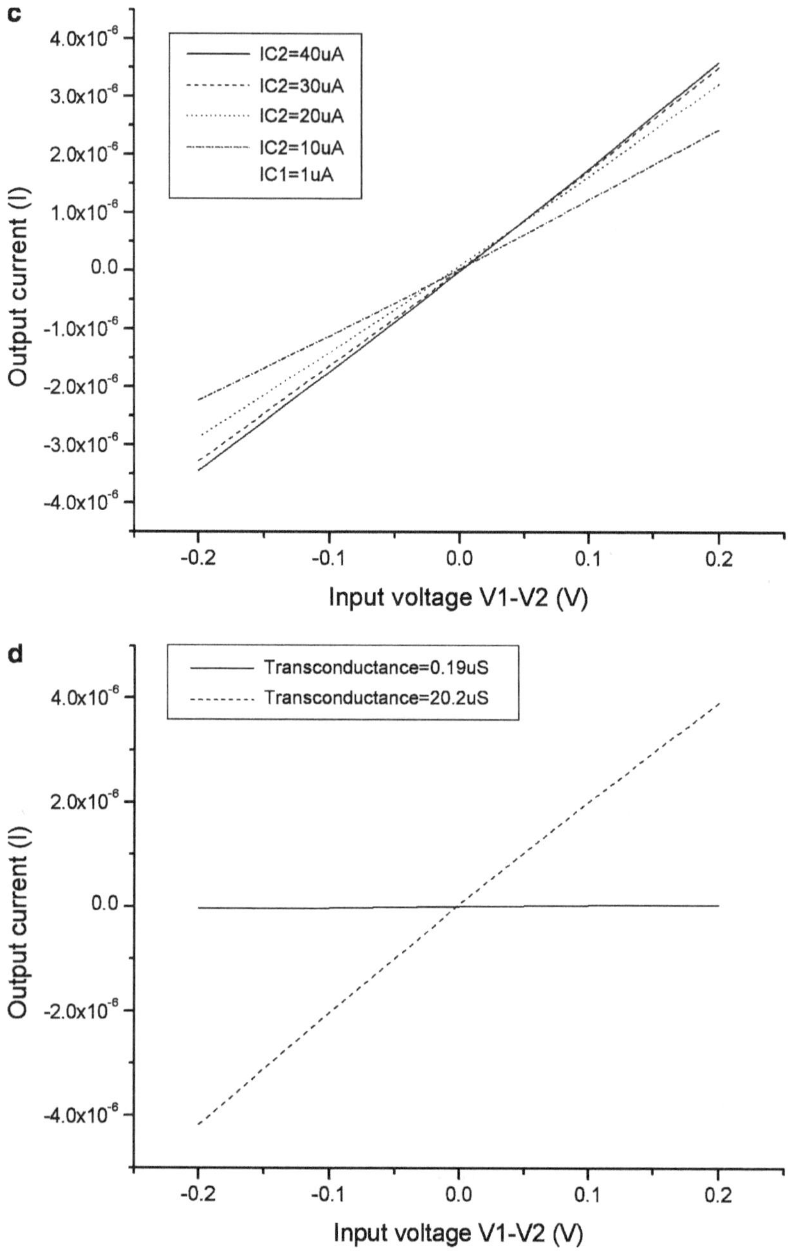

Fig. 3.20 (continued) (**c**) transconductor in the strong inversion region (**d**) transconductor in the multi-inversion regions: the input stage stays in the strong inversion region while the output stage operates from the strong inversion region to the weak inversion region

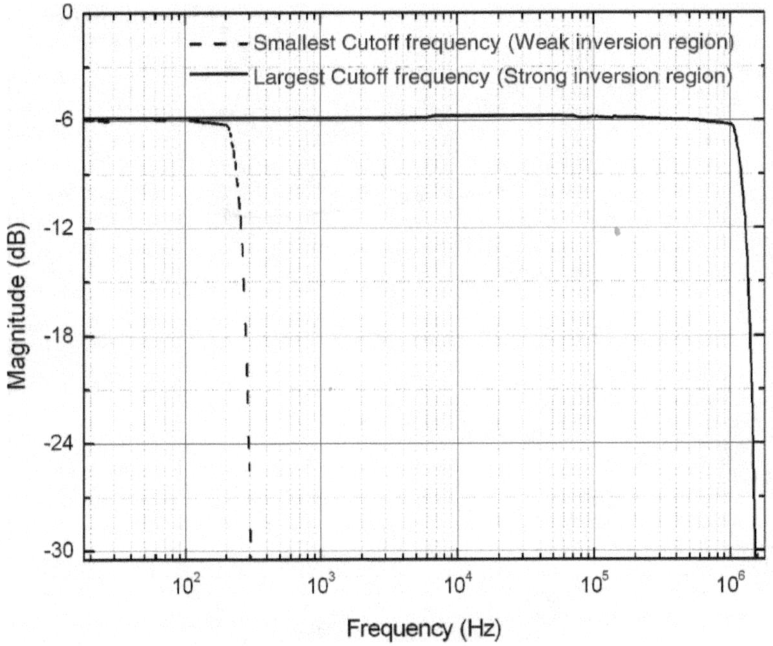

Fig. 3.21 Measured frequency responses over the tuning range

signal. Input signals with higher frequency are applied here owing to the fact that the large transconductance would be used for the filter with higher cutoff frequency. As I_{C2} decreases to 0.5 μA, the output stage of the transconductor operates in the weak inversion region while the input stage stays in the same region, and THD of −42 dB is measured.

Figure 3.21 illustrates the filter measurement results over the tuning range for different bias currents I_{C1} and I_{C2}. The cutoff frequency can be tuned from 250 Hz to 1 MHz, a tuning ratio of 4,000. The third-order inter-modulation distortion, which implies the linearity performance of the proposed filter, is measured at cutoff frequency. By using two sinusoidal tones with the amplitude of 300 mV_{pp}, −53 dB is measured when the filter operates at low cutoff frequency of 250 Hz. On the other hand, when the cutoff frequency of the filter is tuned to 1 MHz, the IM3 is measured −41 dB. Figure 3.22 shows the relationship of measured IM3 versus cutoff frequency for the proposed wide tuning range filter. The worst case which happens in the middle band of the wide tunable filter is due to the multi-inversion operation.

When the cutoff frequency of the filter is 250 Hz, a dynamic range of 52 dB is measured with the 300 mVpp input signal as V_{mode} is set to ground voltage. On the other hand, at the cutoff frequency of 1 MHz, a 48 dB dynamic range is measured as V_{mode} is set to VDD voltage. The measured PSRR at 100 Hz is 36 dB. The filter dissipates 0.2 mW and 0.8 mW for lowest and highest cutoff frequency setting, respectively, at 1.8-V supply.

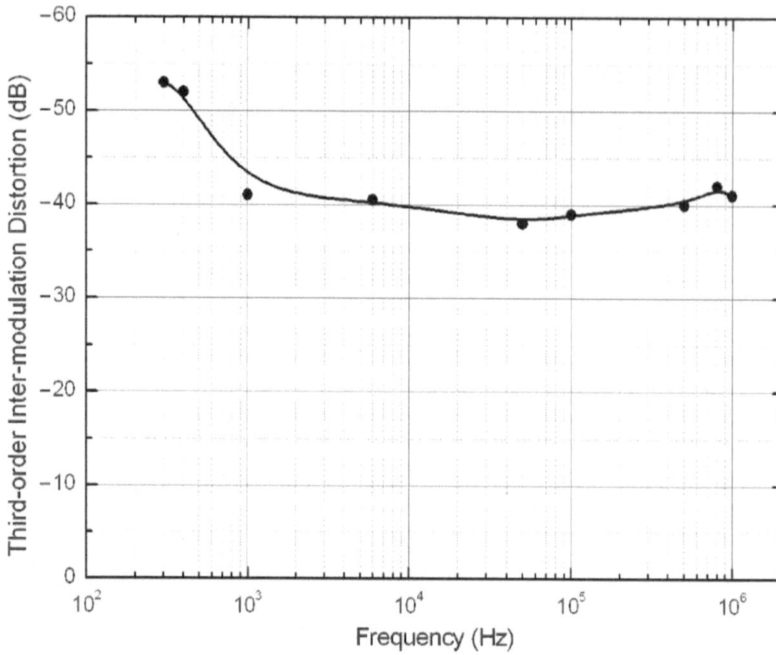

Fig. 3.22 Measured IM3 values at cutoff frequency

Fig. 3.23 Die microphotograph

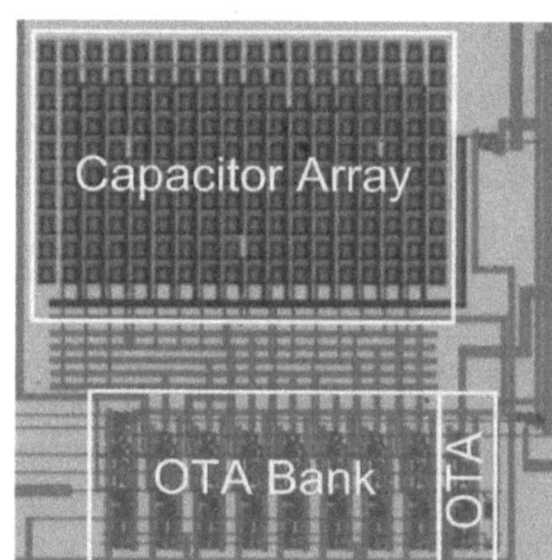

A die photo is shown in Fig. 3.23. The total active area is less than 0.3 mm^2. In order to compare with different implementations of G_m-C low-pass filters, the figure of merit defined in [44], which takes the inter-modulation free dynamic range, speed of the implemented filter, tuning ratio, and normalized power consumption into account, is expressed as follows

$$FoM = 10 \log \left(\frac{IMFDR_{\text{linear}} \times f_o \times tuning}{power_{\text{N}}} \right) \tag{3.91}$$

where IMFDR is defined as the signal to noise ratio when the power of the third-order inter-modulation distortion term equals to the noise power, f_o is the geometrical mean of the cutoff frequency in the unit of Hz, tuning is the tuning ratio of the low-pass filter, and $power_N$ is the geometrical mean of power per pole quantity in the unit of Watt. The FOM of the filter is plotted versus supply voltage and compared with the other previously published works. As shown in Fig. 3.24, our wide tuning range low-pass filter compares favorably with the literature. Table 3.1 summarizes the experimental results of the proposed filter.

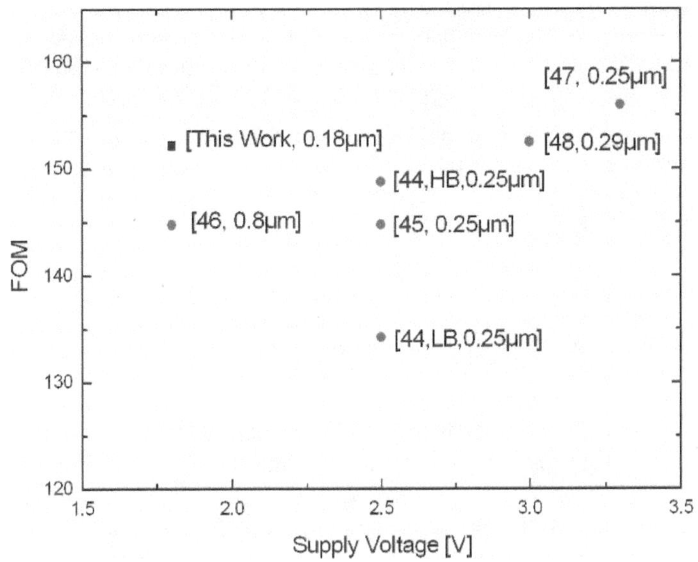

Fig. 3.24 FOM comparison with previously published filters

Table 3.1 Performance summary of the fabricated prototype

Technology	TSMC 180nm CMOS
Supply Voltage	1.8–V
Filter type	Fifth-order Elliptic low-pass
Tuning range	250Hz-1MHz
IM3 for 300mV_{pp} input signals over the tuning range	-40dB
Dynamic range	48 dB
Power consumption	0.8mW
Active area	0.3mm^2

3.3.6 Summary

A CMOS implementation of a fifth-order Elliptic low-pass G_m-C filter with wide tuning range is presented. A complete set of experimental results are provided to demonstrate the validity of the proposed filter. The transconductor used in the filter works in the weak inversion region to achieve micro-power consumption and works in the strong inversion region to extend its tuning range. The working mode of the transconductor could be set and the transconductance could be continuously tuned by setting I_{C1} and I_{C2}. With the use of the transconductor as a building block in the filter architecture, the cutoff frequency of the low-pass filter is changed from 250 Hz to 1 MHz, which covers the range of some audio, speech, bio-medical, and wireless application, so it could be used in multi-mode applications of signal processing. Moreover, the silicon area is saved with less power consumption in the entire filter design. Measurement results demonstrate the potential of the technique to provide a system-on-a-chip design solution for low power dissipation and very wide tuning range applications.

3.3.7 Analysis for Circuit at 1-V Supply

When the supply voltage shown in Fig. 3.15 is reduced to 1-V, one problem is the required headroom for transconductor under the saturation region. Although transistor M9 is not required by giving a suitable CMFB circuit, the margin is still critical, and then the tuning range of the circuit is decreased. Besides, for the equivalent resistor circuit, the signal operating swing at each terminal is reduced to maintain required output conductance. However, a poly resistor with the value of several mega ohms can be used. The other problem is the reduced input signal swing range. The lower boundary of input signal is caused by the tail current source of translinear input stage. The value would not be affected by the reduced supply voltage. The upper boundary of input signal is caused by input source followers, and it has a direct relationship to the variation of supply voltage. Therefore, the proposed transconductor is hard to work at 1-V supply by using 180 nm CMOS process. When the device size is scaled down to 130 nm CMOS process, the benefit of reduced threshold voltage can resist the effect caused by reduced supply voltage. Figure 3.25 shows the simulated V-I relationship of the proposed transconductor in 130 nm CMOS process. Note that in this figure, the transconductor operates under weak inversion region. Figure 3.25a shows the result by using the active equivalent resistor, and a limited input signal range, which is less than 100 mV, is required to maintain linearity. Therefore, the poly resistor is used to replace the active equivalent resistor. The result is shown in Fig. 3.25b and we can find that the linearity can be maintained to be the same as the

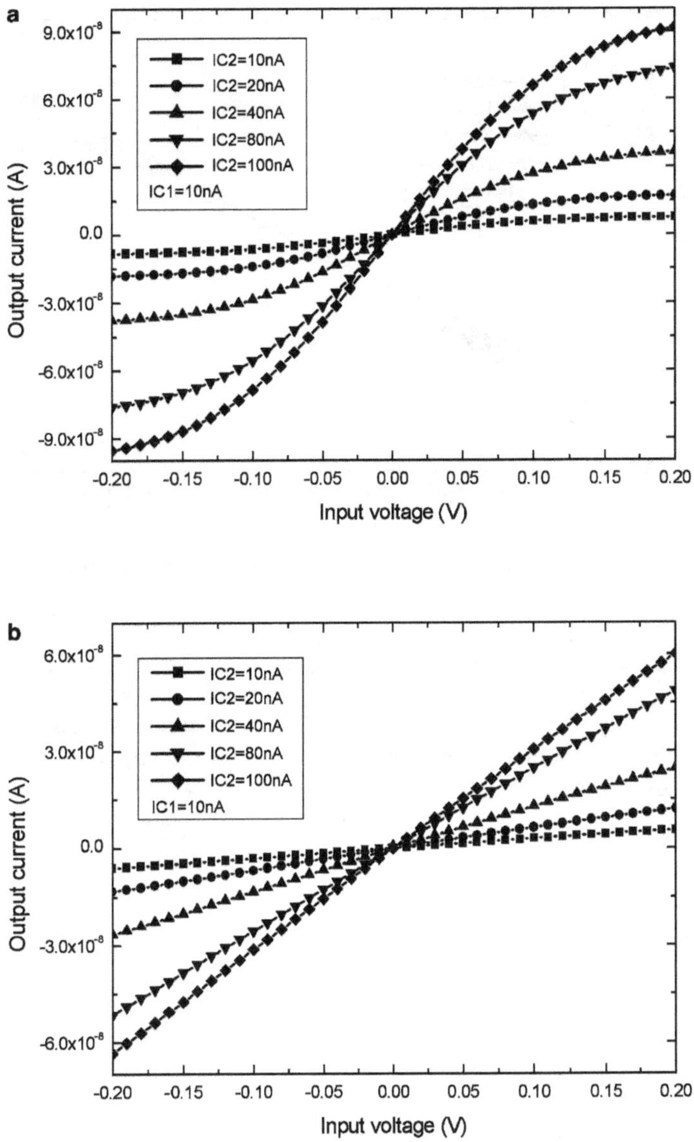

Fig. 3.25 Simulated transconductance tuning range (weak inversion region) at 1-V supply: (**a**) the equivalent active resistor (**b**) passive poly resistor

condition at 1.8-V supply in 180 nm CMOS process. Figure 3.26 shows the result when transconductor works under the saturation region. Required operation region can be maintained owing to smaller threshold voltage. Moreover, larger transconductance is achieved since the gate voltage of the MOSFET resistor is also reduced

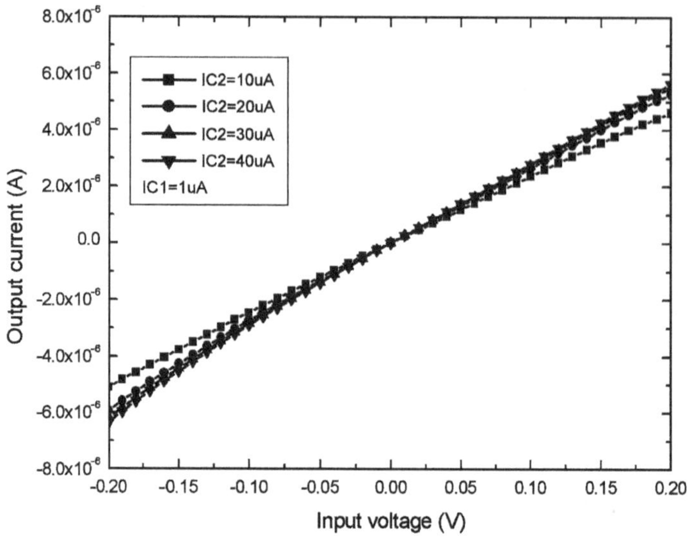

Fig. 3.26 Simulated transconductance tuning range (saturation region) at 1-V supply

by giving a smaller supply voltage. The simulation result shows PSRR of the filter is 36 dB, and it is also the same as the measured result. In addition, since the current is biased under the same condition, smaller power dissipation can be achieved. Therefore, we can conclude the circuit operates well at 1-V supply by using 130 nm CMOS process.

Chapter 4
Multi-mode Channel Selection Filter for Wireless Application

4.1 Introduction

This chapter focuses on the design of the multi-mode wireless channel selection filters, which are designed based on the G_m-C implementation. In this chapter, the zero-IF receiver architecture is discussed at first. Then, the three channel selection filters which meet different wireless and mobile specifications are presented.

4.2 Zero-IF Receiver

As the level of integration in RF transceivers increases, CMOS emerges as the technology with the greatest potential for cost effectiveness. This will be particularly true when wireless communication is integrated in multimedia systems. When designing a wireless receiver, one of the most important parts is the channel filtering to separate the desired signal from the unwanted ones. Recent demand for multi-standard transceivers uses the direct-conversion structure owing to the high integration of a single chip and the ease of system design. Figure 4.1 shows the block diagram of the direct-conversion receiver. It uses a local oscillator to synchronize the carrier frequency of the required signal. Then, the required signal will be selected by a mixer to the band of interest. Since the unwanted signals are still left at the adjacent channel, the received signal will be selected by the channel selection filter for further demodulation. However, the array of channel selection filters for multi-standard applications needs large chip areas, and thus there is a strong motivation to realize the baseband filter in multi-mode applications.

T.-Y. Lo and C.-C. Hung, *1V CMOS* G_m-C *Filters: Design and Applications*, Analog Circuits and Signal Processing, © Springer Science+Business Media B.V. 2009

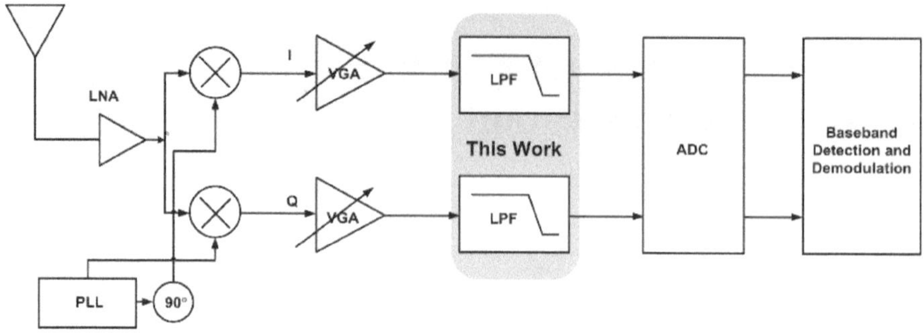

Fig. 4.1 The basic diagram of the direct-conversion receiver

4.3 A G_m-C Continuous-Time Analog Filter for Multi-mode Wireless Applications

A CMOS transconductor for a wide tuning range filter application is presented. The linear transconductor is designed based on the flipped-voltage follower circuit and can work under weak, moderate, and strong inversion regions to maximize the transconductance tuning range. The transconductance tuning can be achieved by changing the bias current of the active resistor. A third-order Butterworth low-pass filter implemented with the transconductors was designed by TSMC 180 nm CMOS process. The results show that the channel selection filter can operate with the cutoff frequency of 500 kHz to 12 MHz. The wide tuning range will be suitable for the specifications of IEEE 802.11a/b/g Wireless LANs, Wideband CDMA, cdma2000, and bluetooth simultaneously under the consideration of saving chip areas. In the design, the maximum power consumption is 14.4 mW with the cutoff frequency of 12 MHz under a 1.8-V supply.

4.3.1 Introduction

Multi-mode wireless applications are required to integrate in a long standby system. Cost efficiency has been greatly increased with the emergence of CMOS technology in high performance VLSI implementations. Moreover, to save the silicon area in a standalone system-on-a-chip solution, re-usable circuits for different system applications can be even cost-effective.

For wireless communication, the channel-selection filter is an important component for many specifications. To meet different specifications for the desired channel, new basic analog building blocks should be re-designed. This work presents the CMOS implementation of a low-pass G_m-C filter for a very wide frequency tuning range for multi-mode wireless applications. The transconductor, which performs the voltage-to-current conversion, is a basic building block in the continuous-time filters. There are numerous works to improve the transconductor linearity, but the reported technology is not suitable for wide tuning range applications. In the section, the high performance linear circuit with the use of an active equivalent resistor is designed based on the flipped-voltage follower structure [49]. The equivalent resistor can operate in the weak, moderate, and strong inversion to achieve a wide transconductance tuning range, and the tuning ability can be achieved by adjusting the bias current. Then, the transconductor is used to design the third-order Butterworth low-pass G_m-C filter, where the designed order is suitable for the desired applications. Section 4.3.2 develops the proposed transconductor architecture. The design of the equivalent resistor and the common-mode feedback circuit is discussed in Section 4.3.3. The G_m-C filter is developed in Section 4.3.4, followed by the simulation results in Section 4.3.5. Summary is presented in Section 4.3.6.

4.3.2 *Proposed Transconductor Circuit*

Figure 4.2a shows the implementation of the transconductor, and the circuit is designed based on the flipped voltage follower (FVF) structure. The FVF structure, composed of transistors M1(M2) and M3(M4), and bias current I_b, is a very efficient low-voltage implementation. Since the follower is biased at the drain terminal rather the source terminal, the circuit behaves as the conventional source follower with even lower output impedance. Due to the shunt feedback through transistor M3(M4), the equivalent impedance at the source of transistor M1(M2) is given by

$$Ro_{eq} = \frac{1}{gm_1 gm_3 ro_1} \tag{4.1}$$

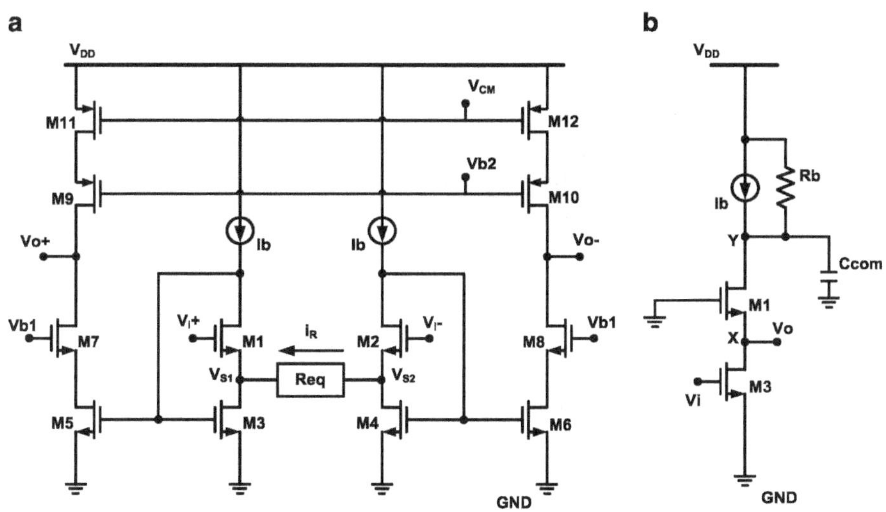

Fig. 4.2 Proposed transconductor circuit: (**a**) implementation of transconductor (**b**) open loop model of the FVF circuit

where parameters gm_i and ro_i are the transconductance and the output resistance of transistor Mi, respectively. The output resistance is a relatively low value compared with the conventional structure, and some of the design constrains can be relaxed under the FVF structure. In the proposed circuit, an equivalent resistor Req is used to perform the voltage-to-current conversion. The input voltage $Vin = V_i + -V_i-$ is applied to the resistor through the very low output impedance buffers. Thus, the negative feedback loop will force $V_i + -V_i- = V_{S1} - V_{Ss}$ to perform the voltage-to-current conversion. We can find that the current $i_R = (V_i + -V_i-)/R_{eq}$, and thus the transconductance would be affected by the value of equivalent resistor. The generated current would flow through transistor M3(M4), and be mirrored by transistor M5(M6) while the drain current ID1(ID2) would be biased to the current source Ib. In addition, the cascode output stage is used to maintain the required gain of the transconductor.

The stability of the transconductor is determined by the stability of the FVF buffers owing to the absence of the other internal nodes, and thus the stability of shunt feedback should be taken into consideration. For the FVF circuit, if we open the loop at the gate of M3 and provide a test voltage source, as illustrated in Fig. 4.2b, the open loop gain of $A_{OL} = -gm_3 R_{OLY}$ would be obtained, where R_{OLY} is the equivalent impedance at node Y. The circuit has a dominant pole at node Y which is equal to $1/R_{OLY}C_Y$ and a non-dominant pole at node X which is equal to $1/R_{OLX}C_X$, where R_{OLX} is the equivalent impedance at node X. For stability issues, the non-dominate pole should be larger than twice the gain bandwidth of the circuit, and thus the constrain $C_X/C_Y < gm_1/4\,gm_3$ should be maintained. The condition is easily achieved by proper sizing of the transistors M1(M2) and M3(M4). Moreover, to further improve circuit stability, a compensating capacitor C_{com} could be also added at the drain node of transistor M1.

4.3.3 The Equivalent Resistor and the CMFB Circuit

4.3.3.1 The Equivalent Resistor

The equivalent resistor can be implemented by the active or passive devices. However, the transconductor should be tuned to meet different specifications under process and temperature variations. For passive devices, we need to turn on and off switches so that different numbers of passive resistors could be selected for the transconductance tuning. However, this programmable solution requires large die area, limits the precision of the filter −3 dB cutoff frequency, and contributes larger noise.

In order to add the tuning ability for the transconductor, active devices, instead of the passive resistors, are used to implement an equivalent resistor with the drawback of degraded linearity performance. Figure 4.3 shows the equivalent resistor circuit. If the transistors MR1 and MR2 are working in the saturation region, we have

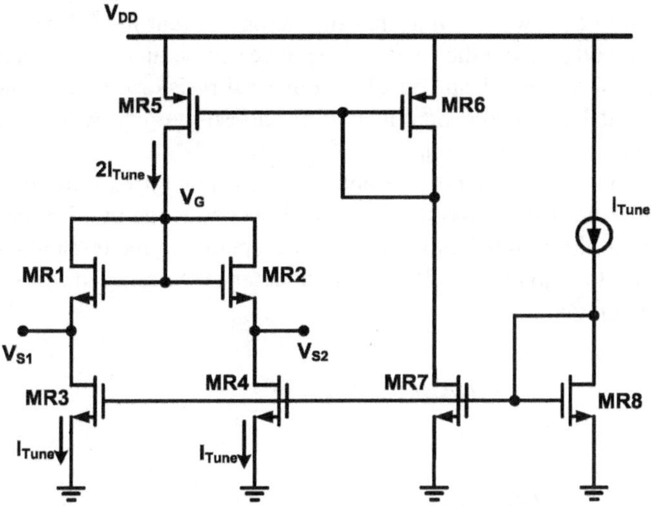

Fig. 4.3 The equivalent resistor circuit

$$i_{R1} = I_{Tune} - i_R = \frac{1}{2}K\left(V_G - V_{S1} - V_{th}\right)^2 \tag{4.2}$$

$$i_{R2} = I_{Tune} + i_R = \frac{1}{2}K\left(V_G - V_{S2} - V_{th}\right)^2 \tag{4.3}$$

where K is the MOS strong inversion parameter and V_{th} is the MOS threshold voltage. By subtracting Equation (4.2) from Equation (4.3), the current i_R through the equivalent resistor is given by

$$i_R = \frac{1}{2}K\left(V_{S1} - V_{S2}\right)\left[2\left(V_G - V_{th}\right) - \left(V_{S1} + V_{S2}\right)\right] \tag{4.4}$$

When we combine the equivalent resistor circuit into the transconductor, as shown in Fig. 4.2a, the DC voltage at nodes V_{S1} and V_{S2} can be defined as

$$V_{S1} = V_{cm} + \frac{V_d}{2} - V_{SF} \tag{4.5}$$

$$V_{S2} = V_{cm} - \frac{V_d}{2} - V_{SF} \tag{4.6}$$

where V_{cm} is the input common-mode voltage, V_d is the input differential-mode voltage, and V_{SF} is the voltage shift from the FVF circuit. By substituting (4.5) and (4.6) into (4.4), we can obtain

$$i_R = K V_d\left(V_G - V_{th} - V_{cm} + V_{SF}\right) \tag{4.7}$$

From Equation (4.7), we can find that the output current has a linear relationship with the input voltage, and the transconductance can be tuned by adjusting the gate voltage of transistors MR1 and MR2. In our final resistor circuit implementation, the bias current I_{TUNE} could be changed to choose the gate voltage and thus the transconductance can also be tuned.

The above design carried out was analyzed based on the assumption that transistors operate in the strong inversion region. However, if the provided current I_{TUNE} is small enough, the equivalent resistor would operate in the moderate or weak inversion region to provide smaller transconductance. Similar analysis can be also performed by giving

$$i_{R1} = I_{Tune} - i_R = I_{D1} \exp\left(\frac{V_G - V_{S1}}{nU_T}\right) \tag{4.8}$$

$$i_{R2} = I_{Tune} + i_R = I_{D1} \exp\left(\frac{V_G - V_{S2}}{nU_T}\right) \tag{4.9}$$

where I_{D1} is the reverse saturation current, n is the weak inversion slope factor, and U_T is the thermal voltage. By subtracting Equation (4.8) from Equation (4.9) and substituting Equations (4.5) and (4.6) into it, the output current which could be approximated through a Taylor series expansion is given by

$$i_R = V_d\left(\frac{1}{nU_T}\right)\left(1 + \frac{V_G - V_{cm} + V_{SF}}{nU_T}\right) \tag{4.10}$$

Again, the linear relationship confirms the circuit operation and thus the proposed transconductor can operate to achieve a wider tuning range for multi-mode wireless applications.

4.3.3.2 The CMFB Circuit

As the transconductor is designed under the differential structure, a common-mode feedback circuit is required to control the output common-mode voltage. For the conventional CMFB circuit shown in Fig. 4.4, the input transistors MF1 to MF4 perform the task of the common-mode detection and reference comparison. If the common-mode voltage at the transconductor output nodes equals to the desired reference voltage, V_{ref}, then the total current through MF7 will be constant and the common-mode bias voltage V_{CM} is fixed. On the other hand, if the common-mode voltage at the transconductor output nodes is not the same as V_{ref}, a current will be mirrored by MF9 to adjust V_{CM} adaptively. Thus, the feedback mechanism makes the output common-mode voltage to the desired value.

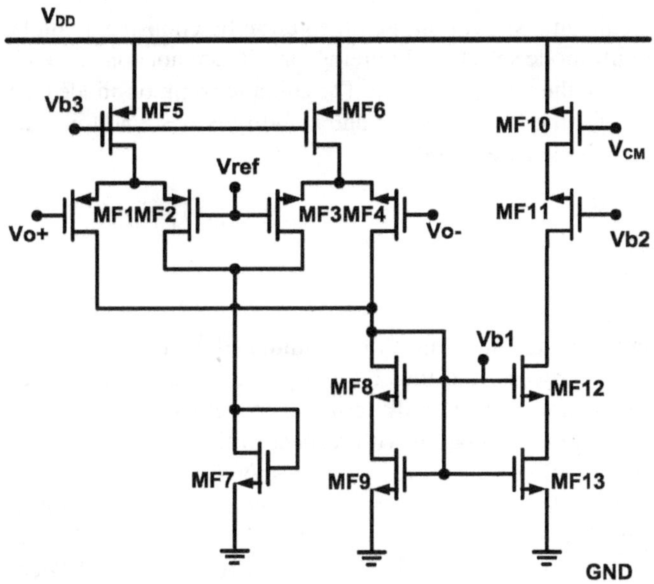

Fig. 4.4 The common-mode feedback circuit

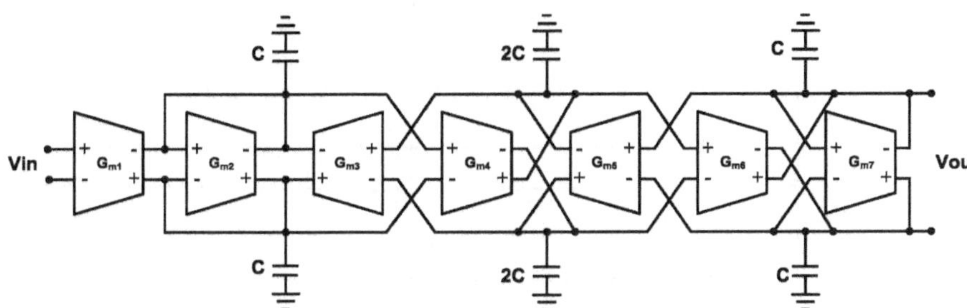

Fig. 4.5 The G_m-C realization of the third-order Butterworth filter

4.3.4 Filter Architecture

To demonstrate the basic building block in a system level, a differential third-order Butterworth low-pass filter is implemented for the required multi-mode strategy. From the demonstration of the passive ladder prototype, the third-order Butterworth low-pass G_m-C filter, which consists of seven identical transconductors, is used as shown in Fig. 4.5. In this low-pass filter design, the cutoff frequency of the filter is proportional to G_m/C, where G_m is the transconductance and C is the capacitance. Each transconductor can be set to a single value so that frequency tuning can be achieved from a single source. Therefore, the cutoff frequency of the G_m-C filter is tuned by changing the bias current of the equivalent resistor in our proposed transconductor. Besides, the tuning circuitry can be further developed with digitally controlled circuits in a system-on-a-chip solution by choosing a number of current sources for multi-mode selection. Q tuning circuits are not considered here with the intrinsic quality of the low Q structure. The suitable common-mode feedback circuit should be used for the filter linearity and stability issues, and thus three common-mode feedback circuits are introduced.

4.3.5 Results

The transconductor and the filter were simulated in the TSMC 180 nm Deep N-WELL CMOS process, and thus the body effects can be simply eliminated by connecting the source and the bulk terminals together in the process. Figure 4.6 shows the large signal simulation with respect to the function of differential input voltage. A tuning ratio of more than 24 corresponding to the adjusted bias current I_{Tune} can be obtained. Figure 4.7 illustrates the filter simulation results at 1.8-V supply. The cutoff frequency can be tuned from 500 kHz to 12 MHz, and the range covers the specifications of IEEE 802.11a/b/g Wireless LANs, Wideband CDMA, cdma2000 and bluetooth. Table 4.1 summarizes the simulation results of this work under different specifications.

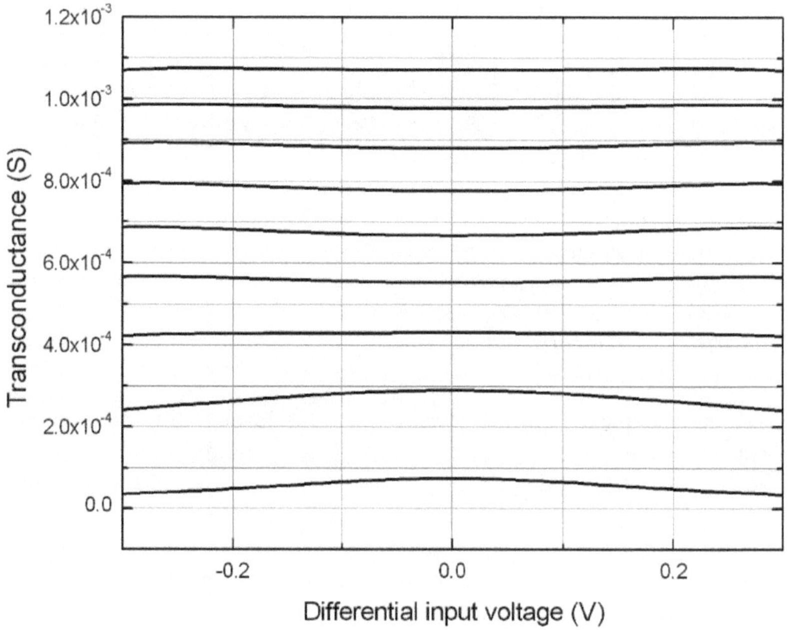

Fig. 4.6 Gm variation of the proposed transconductor

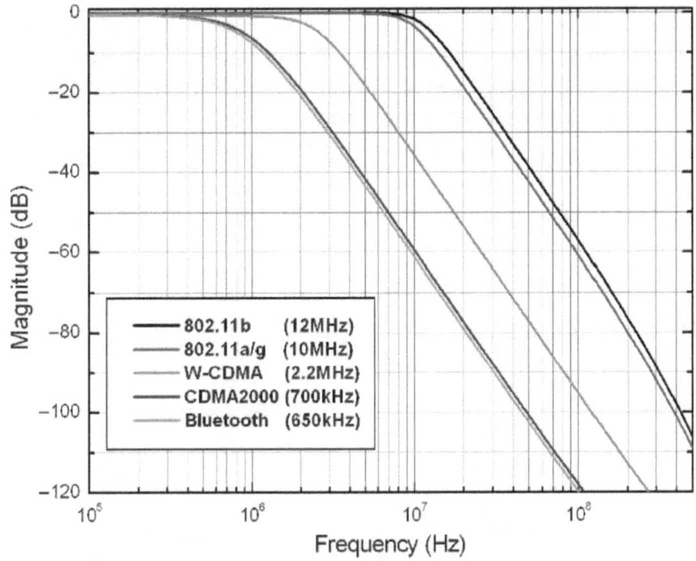

Fig. 4.7 The frequency response of the proposed multi-mode filter

Table 4.1 Performance summary of this work

Technology	180 nm CMOS	
Supply	1.8-V	
Tuning rang	500 kHz – 12 MHz	
IIP3	Bluetooth	23.2 dBm
	CDMA2000	22.7 dBm
	Wideband CDMA	19.3 dBm
	IEEE 802.11a/g	17.5 dBm
	IEEE802.11b	16.9 dBm
Power consumption	Bluetooth	4.96 mW
	CDMA2000	4.98 mW
	Wideband CDMA	5.31 mW
	IEEE 802.11a/g	11.6 mW
	IEEE802.11b	14.4 mW
Output Noise density	180 nV/$\sqrt{\text{Hz}}$ at 2 MHz	

4.3.6 Summary

A CMOS implementation of a third-order Butterworth low-pass G_m-C filter for multi-mode wireless applications is presented. The transconductor is designed based on the FVF structure and the equivalent resistor. Instead of the passive resistor, the active resistor could work in the weak, moderate, and strong inversion regions to extend the transconductance tuning range. With the use of the transconductor as a building block, the cutoff frequency of the low-pass filter can be tuned from 500 kHz to 12 MHz, which could be used for the specifications of IEEE 802.11a/b/g Wireless LANs, Wideband CDMA, cdma2000 and bluetooth. The theoretical analysis shows the high performance operation of the proposed circuit, and a complete set of simulation results is provided to demonstrate the validity of the filter.

4.3.7 Analysis for Circuit at 1-V Supply

When the supply voltage of the circuit used in Fig. 4.2 is reduced to 1-V, the loop gain of the FVF circuit will be degraded. Small loop gain will lead to voltage deviation between input and output followed nodes. Moreover, the drain-to-source margin will also be decreased owing to the cascade structure, and the gain performance is degraded. When the process is scaled down to 130 nm CMOS, the MOSFET with smaller threshold voltage could be used to increase FVF loop gain and the transconductor gain performance. However, to maintain correct operation region for transistors in this circuit, the input signal should be limited within the range of a MOSFET threshold voltage. Due to this constrain, the simulated input signal range by using 130 nm CMOS model, is limited to about 0.2 V for acceptable linearity. The unity-gain bandwidth is 30 MHz and the excess phase is 0.13°. Based on the differential pair fashion, the simulated CMRR is 46 dB. In addition, the equivalent resistor circuit is also affected by the reduced supply voltage. As shown in Fig. 4.3, the upper boundary of the voltage V_G is limited by the supply and drain-to-source voltage of transistor MR5. Since transistor MR5 need to operate at correct region to maintain current mirror function, the upper boundary decreases with direct relationship to supply voltage. However, different from the condition of the FVF circuit, the smaller threshold voltage can extend the lower boundary of the voltage V_G, and thus the effect caused by smaller supply voltage can be relaxed. We should note that the voltage at node V_G directly affects the transconductance, and then the filter cutoff frequency. Figure 4.8 shows the simulated tuning range. Simulation result also shows the tuning range is from 1.5 MHz to 9 MHz by connecting a smaller loading capacitor, which shows a slightly degraded performance compared with 1.8-V supply in 180 nm CMOS process.

4.4 Multi-Mode G_m-C Channel Selection Filter for Mobile Applications

A CMOS transconductor for multi-mode channel selection filter is presented. The transconductor includes a voltage-to-current converter and a current multiplier. Voltage-to-current conversion employs linear region MOS transistors and the circuit

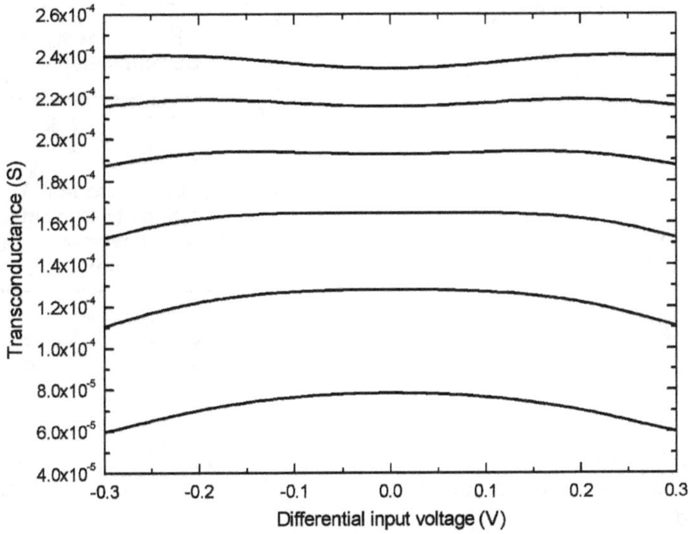

Fig. 4.8 Gm variation of the proposed transconductor at 1-V supply

features high linearity over a wide input swing range. The current multiplier, which operates in the weak inversion region, provides a wide transconductance tuning range without degrading the linearity. A third-order Butterworth low-pass filter implemented with the transconductors was designed by TSMC 180 nm CMOS process. The measurement results show that the filter can operate with the cutoff frequency of 135 kHz to 2.2 MHz. The tuning range and the linearity performance would be suitable for the wireless specifications of GSM, Bluetooth, cdma2000, and Wideband CDMA. In the design, the maximum power consumption at the highest cutoff frequency is 2 mW under a 1-V supply.

4.4.1 Introduction

Recent trend in portable solutions is to include multiple applications in a high-integration system. Therefore, cost efficiency of VLSI implementations has been greatly enhanced with the emergence of multi-mode technology. This work presents a CMOS implementation of a channel selection filter for multi-mode mobile applications. The channel-selection filter is an important component for direct conversion receiver in wireless applications. It acts between the variable gain amplifier and the baseband detection block. A large tuning rang of the low-pass filter would be required for various wireless applications.

To meet different specifications for the desired channel in multi-mode technology, new basic analog building blocks should be re-designed. The transconductor, which performs the voltage-to-current conversion, is a basic building block in the

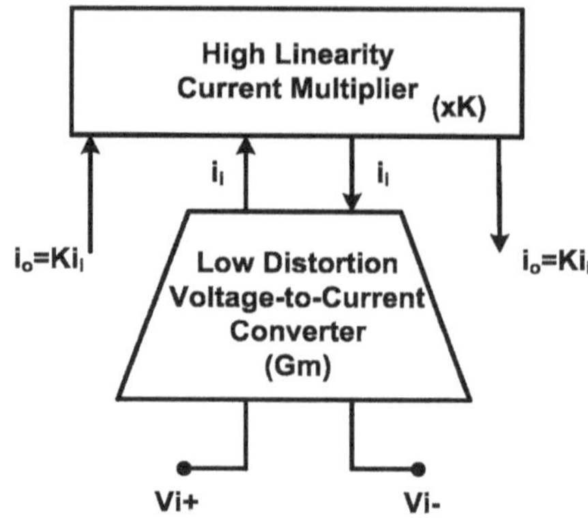

Fig. 4.9 Basic diagram of the transconductor

G_m-C continuous-time filters. For the multi-mode applications, the transconductor requires a wide tuning range of the transconductance while the linearity is well maintained.

In this section, a high performance linear transconductor is designed by the combination of a voltage-to-current converter and a current multiplier. The basic block diagram is shown in Fig. 4.9. The linear region MOS transistors are used to perform the voltage-to-current conversion and the third-order harmonic distortion term can be minimized by providing suitable bias conditions. The current multiplier makes use of weak inversion region MOS transistors to achieve high linearity. Then, the transconductor is used to design a third-order Butterworth low-pass G_m-C filter, where the filter order is suitable for the desired applications. Section 4.4.2 develops the proposed high linearity transconductor. The G_m-C filter, which meets the specifications of GSM, Bluetooth, cdma2000, and Wideband CDMA, is developed with the comments of measurement results in Section 4.4.3. Finally, the results are summarized in Section 4.4.4.

4.4.2 The Proposed Transconductor Circuit

4.4.2.1 The Triode Region MOS Characteristic

The transconductor is designed with a differential input pair operating in the triode region. The drain current of a MOS transistor in the linear region can be expressed as

$$I_D = K_{lin}\left[(V_{GS} - V_{th})V_{DS} - \frac{1}{2}V_{DS}^2\right] \tag{4.11}$$

where K_{lin} is the device parameter, V_{GS} is the gate-to-source voltage of the MOS transistor, V_{DS} is the drain-to-source voltage of the MOS transistor, and V_{th} is the threshold voltage. Thus, the output current will have a linear relationship with the applied input gate-to-source voltage under a constant drain-to-source voltage. The transconductance, obtained from the derivative of the current-to-voltage characteristic, can be expressed as

$$G_m = \frac{\partial I_D}{\partial V_{GS}} = K_{lin} V_{DS} \tag{4.12}$$

The above equation is obtained by assuming that the V_{DS} voltage doesn't change with the variation of V_{GS} voltage. However, efforts should be made to maintain the independence in realistic circuit implementation. Previous research [50] used the cascode structure to decrease the variation of drain voltage, but the circuit required BiCMOS technology and the voltage variation at the source terminal degraded the linearity. Regulated cascode structures have been used with the expense of increased complexity and power consumption [51]. Besides, the pseudo-differential structure was introduced with the need of the CMFF circuit to solve the problem caused by input common-mode variation. Therefore, to design the transconductor based on the triode region input, we need to keep the linearity without large power consumption, and the common-mode problem should be solved.

The transconductance is proportional to the V_{DS} voltage from (4.12), so the transconductance tuning can be achieved. However, if a large tuning range is required, the MOS transistor operation will shift from the triode region to the saturation region owing to the increased V_{DS} voltage and thus degrading the linearity performance.

4.4.2.2 The Transconductor Implementation

Figure 4.10 shows the proposed transconductor circuit. The circuit is designed based on FVF structure, which is composed by transistors M5 and M6. The main voltage-to-current conversion is provided by transistors M1 and M2. The gate voltage of transistor M5 is used to provide a bias voltage for transistors M3 and M4, and make sure the linear region operation of transistors M1 and M2. Thus, the drain voltage of transistors M1 and M2 will be kept at a constant value. Besides, the source of transistors M1 and M2 is fixed to a constant value owing to the FVF feedback loop. From the analysis, it is shown that the low impedance can be obtained at the source of transistor M6. The structure suppresses the variation at the source of transistors M1 and M2, and thus the circuit would operate under a class-AB fashion.

To obtain the voltage-to-current characteristic of our proposed circuit, the input differential voltages at the gate of transistors M1 and M2 are given by

$$V_{i+} = V_{cm} + \frac{v_d}{2} \tag{4.13}$$

$$V_{i-} = V_{cm} - \frac{v_d}{2} \tag{4.14}$$

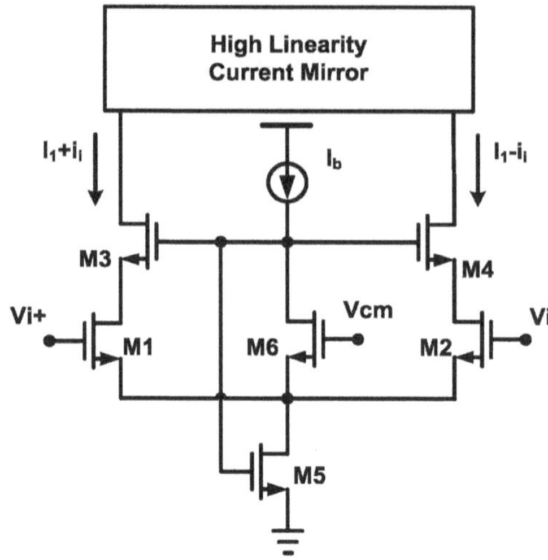

Fig. 4.10 The voltage-to-current cell

where V_{cm} is the input common-mode voltage and v_d is the input differential-mode voltage. The currents flowing through transistors M1, M2, and M6 are expressed by

$$I_{D1} = K_1\left[(V_{cm} + \frac{v_d}{2} - V_{th})(V_{D1} - V_X) - \frac{1}{2}(V_{D1} - V_X)^2\right] \tag{4.15}$$

$$I_{D2} = K_2\left[(V_{cm} - \frac{v_d}{2} - V_{th})(V_{D2} - V_X) - \frac{1}{2}(V_{D2} - V_X)^2\right] \tag{4.16}$$

$$I_b = \frac{1}{2}\left[K_6(V_{cm} - V_X - V_{th})^2\right] \tag{4.17}$$

where V_{Di} is the drain voltage of transistor M_i and V_x is the source voltage of transistor M6. The aspect ratio of transistors M1, M2, and M6 are set to the same value. To analyze the linearity of the output current against the input voltage under the differential structure, a Taylor series expansion is used and then the relationship becomes

$$i_{out} = a_{1,out}v_d + a_{2,out}v_d^2 + a_{3,out}v_d^3 + a_{4,out}v_d^4 + \cdots \tag{4.18}$$

$$a_{1,out} = G_m = \frac{1}{2}K\left(\alpha - \frac{\beta^2 - 3\beta\alpha + \alpha^2}{\sqrt{\alpha^2 + 2\beta\alpha - \beta^2}}\right) \tag{4.19}$$

$$a_{3,out} = -\frac{3(\beta - 2\alpha)^3\alpha}{2(\beta - \alpha)(\beta^2 - 2\alpha\beta - \alpha^2)^3} \tag{4.20}$$

$$\text{where}\quad K_1 = K_2 = K, \quad \alpha = \sqrt{\frac{2I_b}{K_6}}, \text{ and } \beta = V_{cm} - V_{th}$$

In (4.18), the output current seems a nonlinear function of the input differential voltage. Since the even-order harmonic terms cancels out by the differential structure, the third-order harmonic distortion becomes the dominant component. We can find that to minimize the third distortion term, the following equation should be satisfied.

$$\sqrt{\frac{2I_b}{K_6}} = \frac{V_{cm} - V_{th}}{2} \tag{4.21}$$

If the above condition is held, only the fifth or higher order distortion terms are left in (4.18) and these nonlinear terms have very low contribution to the proposed circuit.

For the problem of input common-mode variation, we do not need the CMFF circuit, which would consume more power. This is because when the gate terminal of transistor M6 is biased at the common-mode voltage V_{cm}, we can have very low input common-mode gain and then high CMRR can be obtained.

4.4.2.3 The High Linearity Current Multiplier

Low distortion voltage-to-current conversion is obtained under the defined bias conditions, and furthermore, the transconductance tuning ability could be added with the high linearity current multiplier circuit, as illustrated in Fig. 4.11. The transistors operate in the weak inversion region while proper sizing is required. For a MOSFET operating in the weak inversion region with V_{DS} larger than a few times of ther-

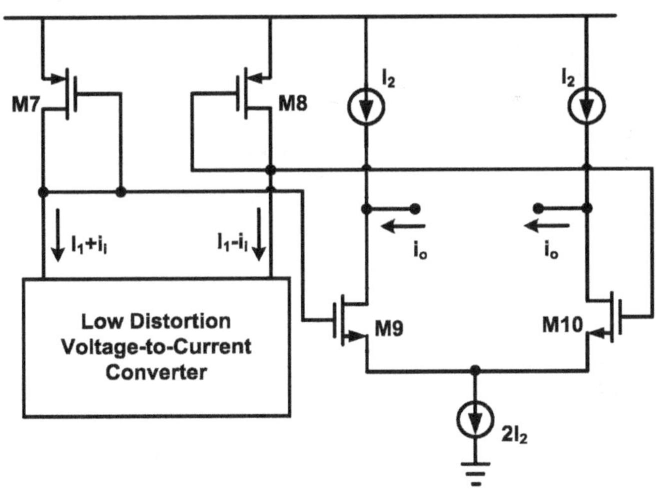

Fig. 4.11 The high linearity current multiplier

mal voltage U_T, its current exhibits an exponential dependence of V_{GS} and can be expressed as

$$I_D = I_{D0}\frac{W}{L}\exp\left(\frac{V_{GS}}{nU_T}\right) \tag{4.22}$$

where W and L are the width and the length of the transistor, respectively, I_{D0} is the reverse saturation current, n is the weak inversion slope factor, and U_T is the thermal voltage. From the above equation, we can find that

$$V_{SG7} = nU_T \ln\left(\frac{I_{D7}}{I_{D0}}\frac{L_7}{W_7}\right) \tag{4.23}$$

$$V_{SG8} = nU_T \ln\left(\frac{I_{D8}}{I_{D0}}\frac{L_8}{W_8}\right) \tag{4.24}$$

The device sizes of transistors M7 and M8 are set to the same. From the linear conversion, we can obtain $I_{D7} = I_1 + i_i$ and $I_{D8} = I_1 - i_i$. We can have

$$\Delta V = V_{SG7} - V_{SG8} = V_{GS9} - V_{GS10} = nU_T \ln\left(\frac{I_1 + i_i}{I_1 - i_i}\right) \tag{4.25}$$

The weak inversion slope factor n is equal to $(C_{ox} + C_{depl})/C_{ox}$ [51], where C_{ox} and C_{depl} are the gate and depletion capacitance per unit area, respectively. We can find that the factor is almost equal between NMOS and PMOS devices from the process device model. Thus, the current output from the differential pair of transistors M9 and M10 can be expressed as

$$I_{D9} = \frac{2I_2}{1 + e^{\frac{\Delta V}{nU_T}}} = \frac{2I_2}{1 + \left(\frac{I_1+i_i}{I_1-i_i}\right)} = I_2 - \frac{I_2}{I_1}i_i \tag{4.26}$$

$$I_{D10} = \frac{2I_2}{1 + e^{\frac{-\Delta V}{nU_T}}} = \frac{2I_2}{1 + \left(\frac{I_1-i_i}{I_1+i_i}\right)} = I_2 + \frac{I_2}{I_1}i_i \tag{4.27}$$

Finally, the output current i_o is given by

$$i_o = I_{D10} - I_2 = I_2 - I_{D9} = \frac{I_2}{I_1}i_i \tag{4.28}$$

With the weak inversion characteristic, the input voltage should be logarithmically determined at first and through the exponential function, the output current would have a linear relationship. In other words, the output current is equal to a scaled version of the input current where the scaling factor is determined by the ratio of two bias currents. In our proposed circuit, the output current can be tuned by tuning the bias current I_2 at the output stage. The high linearity characteristic can be also maintained when the circuit operates from the weak inversion region to the moderate

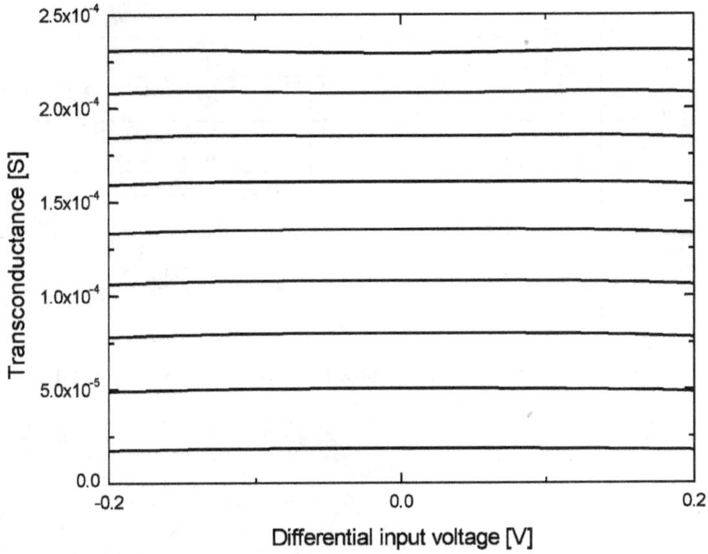

Fig. 4.12 The simulated Gm range of the proposed transconductor

inversion region, and thus the tuning range could be further extended. Figure 4.12 shows the large signal simulation of the transconductance with respect to the function of applied differential input voltage. With the scalable tuning current, the tuning range of 10–230 μS can be obtained.

4.4.2.4 The CMFB Circuit

As the transconductor is designed under the differential structure, a common-mode feedback circuit is required to control the output common-mode voltage. The final circuit implementation, which includes the CMFB circuit, is shown in Fig. 4.13. The aspect ratio of transistors M19 will be twice the value of transistors MF1 and MF2. The input transistors MF3 to MF6 perform the task of the common-mode detection and reference comparison, and the feedback mechanism makes the output common-mode voltage to the desired value.

4.4.3 Filter Architecture and Measurement Result

From the demonstration of the passive ladder prototype, the third-order Butterworth low-pass G_m-C filter, which consists of seven identical transconductors, is chosen as shown in Fig. 4.5. Then, the frequency tuning is achieved by using a single source, I_{tune}.

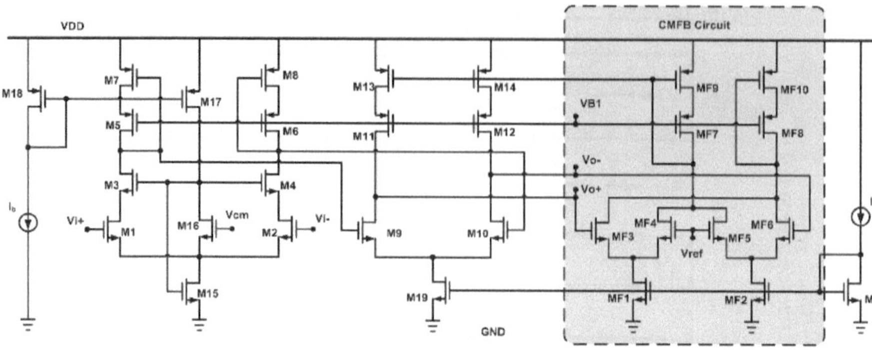

Fig. 4.13 The final implementation of the proposed transconductor with the CMFB circuit

Fig. 4.14 The chip micrograph

The transconductor and the filter were designed in the TSMC 180 nm CMOS process. The chip micrograph is shown in Fig. 4.14 with the active area less than 0.5 mm^2. Figure 4.15 illustrates the filter frequency response at 1-V supply. We should note that the magnitude is normalized owing to the use of the output buffer. The cutoff frequency can be tuned from 135 kHz to 2.2 MHz, and the range covers the specifications of GSM, Bluetooth, cdma2000, and Wideband CDMA. The linearity is maintained by holding the condition of (4.21), and the V_{cm} is set to 0.6 V. The IIP3 is measured by applying a signal with two sinusoid tones near the selected cutoff frequency.

The linearity performance of 23.1 dBm to 19.3 dBm for different specification is obtained. Owing to the weak inversion operation, the small power consumption of 1.57–1.92 mW at different cutoff frequencies is measured for the proposed filter. The measured CMRR of 47 dB is obtained at 50 kHz. Table 4.2 summarizes the measurement results of this work.

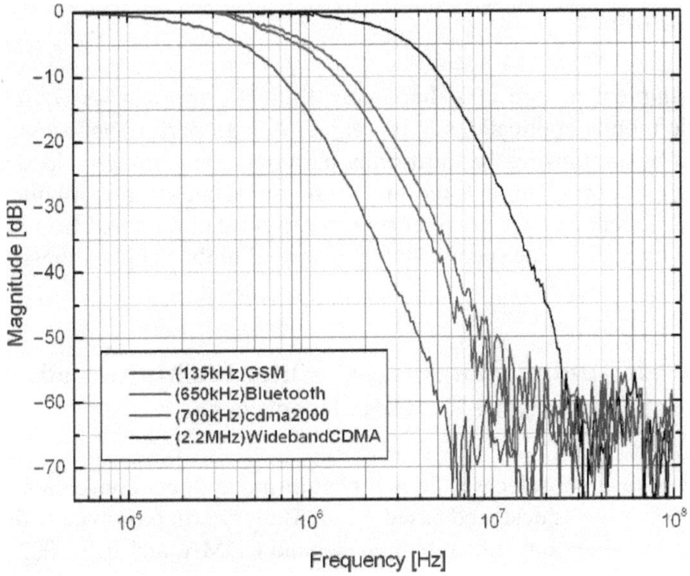

Fig. 4.15 The measured frequency responses of the proposed multi-mode filter

Table 4.2 Performance summary of this work

Technology	180 nm CMOS	
Supply	1-V	
Active Area	<0.5 mm^2	
Tuning range	135 kHz – 2.2 MHz	
IIP3	GSM	23.1 dBm
	Bluetooth	21.2 dBm
	cdma2000	21.0 dBm
	Wideband CDMA	19.3 dBm
Power consumption	GSM	1.57 mW
	Bluetooth	1.72 mW
	cdma2000	1.73 mW
	Wideband CDMA	1.92 mW
Output Noise density	65 nV/$\sqrt{\text{Hz}}$ at 1 MHz	

4.4.4 Summary

A CMOS implementation of a third-order Butterworth low-pass G_m-C filter for multi-mode mobile applications is presented. A high performance transconductor is designed by minimizing the third-order harmonic tone through selecting suitable designing conditions. Through the use of the transconductor as a building block, the cutoff frequency of the channel selection filter can be tuned from 135 kHz to 2.2 MHz, which meets the specifications of several mobile applications.

4.5 A Wide Tuning Range G_m-C Filter for Multi-mode Direct-Conversion Wireless Receivers

A third-order channel selection filter for multi-mode direct-conversion receiver is presented. The filter is designed based on the Butterworth prototype with the target applications of Bluetooth, cdma2000, Wideband CDMA, and IEEE 802.11 a/b/g/n wireless LANs. Linear-region MOS transistors are used to perform the voltage-to-current conversion. A wide tuning range can be achieved by the current multiplier following the linear voltage-to-current converter. Implemented in the TSMC 180 nm CMOS process, the measurement results show that the filter can operate with the cutoff frequencies in the range of 500 kHz to 20 MHz, and thus meet the requirement of different wireless applications. In the design, the power consumption is 4.1–11.1 mW for minimum and maximum cutoff frequency under a 1.2-V supply. The circuit performance is favorably compared with those reported in the literature.

4.5.1 Introduction

The channel selection filter is designed for direct-conversion structure between the down-conversion mixer and the analog-to-digital converter. The filter is designed based on the G_m-C topology, and has a wide continuous tuning range. The large tuning range can be designed to meet the wireless specifications of Bluetooth (650 kHz), cdma2000 (700 kHz), Wideband CDMA (2.2 MHz), IEEE 802.11a/g (10 MHz), IEEE 802.11b (12 MHz), and IEEE 802.11n (20 MHz) wireless LANs applications. These specifications cover the frequency range from 650 kHz to 20 MHz.

For the G_m-C topology, the transconductor will be used as a basic building block, and the transconductance will be proportional to the −3 dB cutoff frequency of the filter. Thus, when the process and temperature variation is taken into consideration, a tuning ratio of more than 50 should be designed while the linearity performance is maintained.

Section 4.5.2 develops the proposed high linearity transconductor with a wide transconductance tuning range. The third-order Butterworth G_m-C filter which meets the required specification with comments from the measurement results is discussed in Section 4.5.3. Finally, the results are summarized in Section 4.5.4.

4.5.2 *The Proposed Transconductor Circuit*

The main function of a transconductor is to convert the input voltage into the output current with a linear transformation factor, and the transconductor employed in filters must be linear over the input signal swing range. On the other hand, the transconductance should be tuned to compensate for the process and temperature variation of G_m-C topology, and we can model it as a voltage controlled current source. In the circuit implementation, bipolar transistors offer a wide transconductance tenability range because the collector current can be varied with little change in the base-emitter voltage. In contrast, the bias of the MOS transistors should be varied significantly and the supply voltage limits the tuning range, which implies the requirement of a higher supply voltage. Thus, a high linearity and wide tuning range transconductor under the low voltage CMOS technology would not only be needed for multi-mode wireless applications but also be helpful to combine with digital circuits for a system-on-a-chip strategy.

4.5.2.1 The Voltage-to-Current Conversion in CMOS Technology

Figure 4.16a shows a conventional transconductor, where the voltage-to-current conversion is obtained by using an operational amplifier through a passive resistor. The current is then sensed and mirrored to the output node. In this circuit, the

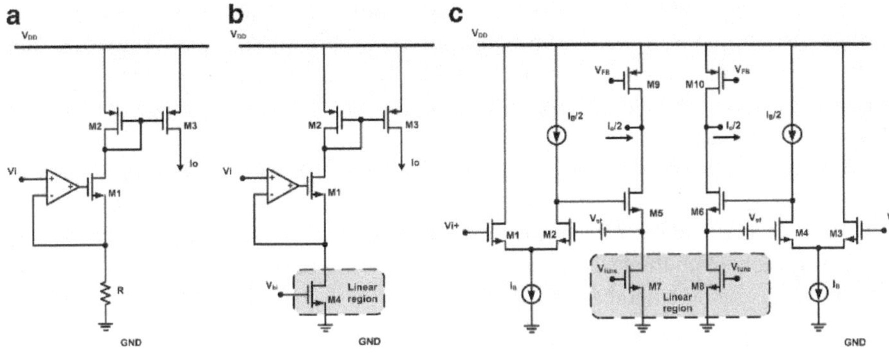

Fig. 4.16 Implementation of transconductor: (**a**) the conventional transconductor (**b**) the MOSFET-only transconductor (**c**) the differential transconductor

array of resistors should be designed to give transconductance tuning ability. This method will increase the chip area, and the required area is highly dependent on the device model. Besides, the precision of the transconductance will be limited through the choice of the programmable resistors. To achieve the transconductance tuning, a single linear region transistor, shown in Fig. 4.16b, is used to replace the passive resistor. This circuit is a MOSFET-only configuration. When the NMOS operates in the linear region, the drain current can be given, and the voltage-to-current relationship in Fig. 4.16b can be obtained. From the simplification of a BISM Level 3 model, the drain current is given by

$$\begin{aligned} I_{D,\,lin} &= K_{lin}(V_{GS} - V_{thn})V_{DS} - \frac{1}{2}\alpha K_{lin} V_{DS}^2 \\ where\ \alpha &= 1 + \frac{1}{2}\gamma\left(2\Phi_f + V_{BS}\right)^{-1/2} \end{aligned} \tag{4.29}$$

where γ is the body effect factor and Φ_f is the surface inversion potential. $K_{lin} = \mu_n C_{ox}(W/L)$, in which W and L are the width and length of the device, respectively, C_{ox} is the oxide capacitance per unit channel area, μ_n is the low-field mobility, and V_{thn} is the NMOS threshold voltage. V_{GS} and V_{DS} are the gate-to-source and drain-to-source voltage, respectively. By taking the process and temperature variation into consideration, α can be chosen to be a value of 1.2.

We can find that the output current would not hold a linear relationship to the input voltage owing to the additional second term in (4.29), which thus degrades the linearity of the transconductor. We should note that the second term in (4.29) forms a square-law equation. Therefore, previous research [52] reports that another transistor can be added to cancel out the second term in (4.29) based on the large signal square-law equation in the saturation region. Thus, the output current will be proportional to the input voltage, and the transconductance can be tuned by adjusting the bias voltage at the gate terminal. In the circuit, a −40 dB THD was reported. There are several disadvantages of the circuit. First, an extra complex circuit needs to be included for current cancellation, and this technique needs an extra operational amplifier, which implies more power consumption. Secondly, the constraints of linear region operation, $V_{GS} - V_t > V_{DS}$, should be hold and this condition is hard to sustain under our wide tuning requirement. Finally, it would not work well in the modern nano-scale CMOS technology owing to that the short channel effect would affect the ideal characteristic of the square-law behavior and then degrade the performance of cancellation ability.

Although the technique reported in [52] is not suitable to provide a high linearity and low power transconductor in modern CMOS process, we can take the circuit in Fig. 4.16b to become the differential version as shown in Fig. 4.16c. To obtain the voltage-to-current characteristic of the circuit, the input differential voltages at the gate of transistors M1 and M2, are given by

$$V_{i+} = V_{cm} + \frac{v_d}{2} \tag{4.30}$$

$$V_{i-} = V_{cm} - \frac{v_d}{2} \tag{4.31}$$

where V_{cm} is the input common-mode voltage and V_d is the input differential voltage. The gate terminals of the linear-region transistors M7 and M8 will be biased at an appropriate voltage, V_{tune}, to make sure the linear region operation. The second term in (4.29) can be cancelled out by inherent differential structure, and thus the output current becomes

$$i_{out} = K_{lin}(V_b - V_{thn} - \alpha V_{cm})v_d \tag{4.32}$$

The equation shows that the linear voltage-to-current conversion can be obtained, and the output current will be dependent on the bias voltage and the input common-mode voltage.

4.5.2.2 The Proposed Transconductor with Tuning Scheme

Followed by the concept of the differential structure in Fig. 4.16c, a wide range transconductance tuning is hard to achieve by adjusting V_{tune} and V_{cm} voltages. In general, V_{tune} should be biased at a higher voltage, such as the supply rail, and then the transistor would always work in the deep triode region to perform highly linear operations. Thus, a high linearity current multiplier would be required following the voltage-to-current conversion.

Figure 4.17 shows the concept of the transconductor, which has a wide transconductance range. In the proposed transconductor, the linear voltage-to-current conversion will be performed at first, and then the transconductance can be tuned by the followed translinear loop. Through the use of feedback loop, linear voltage-to-current conversion can be obtained owing to the fact that the input voltage would be equal to the drain voltage of transistors M1 and M2. Thus, the current flowing through resistors R_1 will have a very high linearity relationship with the input voltage. Then, the transistors M1 to M4 will operate in the weak inversion region while suitable device size is designed. For a MOSFET operating in the weak inversion region with V_{DS} larger than a few times of thermal voltage U_T, its current exhibits an exponential dependence of V_{GS}, and can be expressed as

$$I_D = I_{D0}\frac{W}{L}\exp\left(\frac{V_{GS}}{nU_T}\right) \tag{4.33}$$

where W and L are the width and the length of the transistor, respectively, I_{D0} is the reverse saturation current, n is the weak inversion slope factor, and U_T is the thermal voltage. From the above equation, we can find that

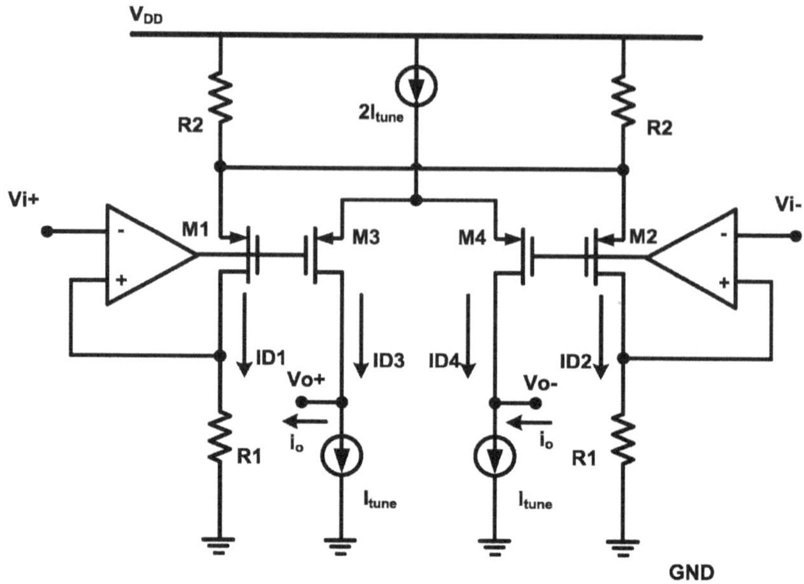

Fig. 4.17 The concept of a wide tuning range transconductor

$$V_{SG1} = nU_T \ln\left(\frac{I_{D1}}{I_{D0}}\frac{L_1}{W_1}\right) \tag{4.34}$$

$$V_{SG2} = nU_T \ln\left(\frac{I_{D2}}{I_{D0}}\frac{L_2}{W_2}\right) \tag{4.35}$$

The device sizes of transistors M1 and M2 are set to the same, and so are transistors M3 and M4. From the linear conversion, we have $I_{D1} = I_1 + i_i$ and $I_{D2} = I_1 - i_i$, where $I_1 = V_{cm}/R_1$. We can have

$$\Delta V = V_{SG1} - V_{SG2} = V_{SG3} - V_{SG4} = nU_T \ln\left(\frac{I_1 + i_i}{I_1 - i_i}\right) \tag{4.36}$$

The current output from the differential pair of transistors M3 and M4 can be expressed as

$$I_{D3} = \frac{2I_{tune}}{1 + e^{\frac{-\Delta V}{nU_T}}} = \frac{2I_{tune}}{1 + \left(\frac{I_1 - i_i}{I_1 + i_i}\right)} = I_{tune} + \frac{I_{tune}}{I_1} i_i \tag{4.37}$$

$$I_{D4} = \frac{2I_{tune}}{1 + e^{\frac{\Delta V}{nU_T}}} = \frac{2I_{tune}}{1 + \left(\frac{I_1 + i_i}{I_1 - i_i}\right)} = I_{tune} - \frac{I_{tune}}{I_1} i_i \tag{4.38}$$

Finally, the output current i_o is given by

$$i_o = I_{D3} - I_{tune} = I_{tune} - I_{D4} = \frac{I_{tune}}{I_1} i_i \tag{4.39}$$

Thus, the output current is equal to a scaled value of the input current. The transconductance tuning can be achieved and the scaling can be determined by the bias current I_{tune}. Moreover, if the transistors are extended from the weak inversion region to the moderate inversion region, the translinear loop would hold well, and an extended tuning range could be obtained. The scaled output current can be combined with the highly liner voltage-to-current conversion, and then the proposed wide tuning range linear transconductor can be achieved. In the circuit, the transconductance in the weak and the moderate inversion region can be expressed as

$$G_{m,weak} = \frac{I_{tune}}{V_{cm}} \tag{4.40}$$

In case of a large I_{tune} current, transistors M13 and M14 will enter into the saturation region. From the equation of a differential amplifier in the saturation region, the output current i_o can be expressed as

$$i_o = \frac{I_{D3} - I_{D4}}{2} = \frac{K_{3,4}}{4} (\Delta V) \sqrt{\frac{4I_{tune}}{K_{3,4}} - \Delta V^2} \tag{4.41}$$

where ΔV is the gate differential voltage as shown in (4.26), and the value can be simplified by neglecting high order terms from a Taylor series expansion. Therefore, we have

$$\Delta V = nU_T \ln\left(\frac{I_1 + i_i}{I_1 - i_i}\right) \approx nU_T \frac{2i_i}{I_1} \tag{4.42}$$

We can substitute (4.42) into (4.41) to obtain the output current. For a small ΔV^2 compared to $4\, I_{tune}/K_{3,4}$

$$i_o \approx \frac{K_{3,4}}{4} (\Delta V) \sqrt{\frac{4I_{tune}}{K_{3,4}}} = nU_T \frac{i_i}{2I_1} \sqrt{K_{3,4} I_{tune}} \tag{4.43}$$

Then, the transconductance becomes

$$G_{m,sat} = nU_T \frac{\sqrt{K_{3,4} I_{tune}}}{V_{cm}} \tag{4.44}$$

From the above equation, the circuit will operate well by holding the linearity performance. We should note that the transconductance would be dependent on the square value of I_{tune} at this condition, rather than the scaled fashion shown in (4.40).

4.5.2.3 The Final Circuit Implementation

Figure 4.18 shows the final circuit implementation of the proposed transconductor. The linear region transistors are used as the resistor R_1 in Fig. 4.17. Unfortunately, the short channel effects still occurs and high order nonlinearity components degrade the linearity of the V-I conversion, especially for nano-scale technology because (4.29) is obtained from the analysis of an approximation. Thus, the distortion components should be analyzed and a new design methodology should be applied to perform a high linearity conversion. For a linear region transistor, the precise transistor model which takes the mobility effect into consideration is expressed as [53]

$$I_{D,\,lin} = K_{lin}\left[\frac{(V_{GS}-V_{thn})\,V_{DS}-\frac{1}{2}\alpha V_{DS}^2}{1+\theta\,(V_{GS}-V_{thn})}\right]\times \left\{1+\frac{R_{eq}K_{lin}}{V_{DS}}\left[\frac{(V_{GS}-V_{thn})\,V_{DS}-\frac{1}{2}\alpha V_{DS}^2}{1+\theta\,(V_{GS}-V_{thn})}\right]\right\}^{-1} \tag{4.45}$$

where θ is the mobility reduction coefficient and R_{eq} is the default device equivalent resistor in the linear region. In the equation, the value of α would be set to one for simplification. By giving $V_{GS,13} = V_{GS,14} = V_{DD}$, $V_{DS,14} = V_{cm} + v_d/2$, and $V_{DS,13} = V_{cm} - v_d/2$, we can derive the voltage-to-current characteristic of MOS transistors M13 and M14. To analyze the linearity of the V_{DS} voltage against the drain current, a Taylor series expansion is used and then the relationship becomes

$$\begin{aligned} I_{D,\,lin} &= I_{D14} - I_{D13} \\ &= a_{1,\,lin}v_d + a_{2,\,lin}v_d^2 + a_{3,\,lin}v_d^3 + a_{4,\,lin}v_d^4 + \cdots \end{aligned} \tag{4.46}$$

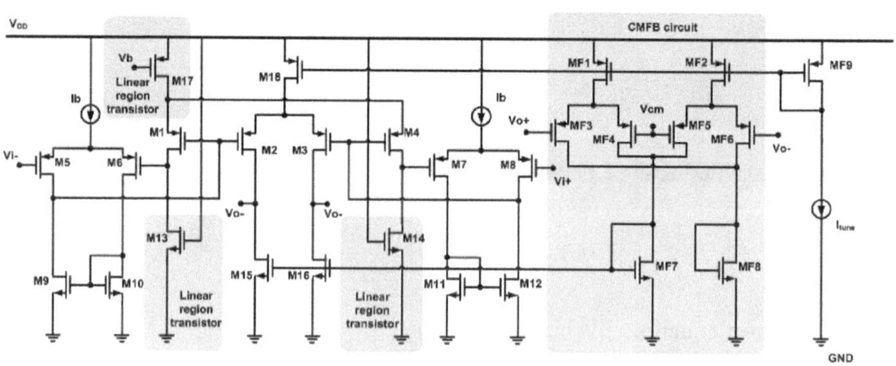

Fig. 4.18 Final implementation of the proposed transconductor with the CMFB circuit

where

$$a_{1,\,lin} = \frac{K_{lin}\left\{R_{eq}K_{lin}V_{cm}^2 + 4\left(V_{DD}-V_{th}-V_{cm}\right)\left[1+\left(R_{eq}K_{lin}+\theta\right)\left(V_{DD}-V_{th}\right)\right]\right\}}{\left\{2+R_{eq}K_{lin}\left[2\left(V_{DD}-V_{th}\right)-V_{cm}\right]+2(V_{DD}-V_{th})\theta\right\}^2}$$

$$a_{3,\,lin} = \frac{-4R_{eq}K_{lin}^2\left[1+\theta\left(V_{DD}-V_{th}\right)\right]\left[1+\left(R_{eq}K_{lin}+\theta\right)\left(V_{DD}-V_{th}\right)\right]}{\left\{2+R_{eq}K_{lin}\left[2\left(V_{DD}-V_{th}\right)-V_{cm}\right]+2(V_{DD}-V_{th})\theta\right\}^4}$$

Owing to the non-ideal characteristic of the linear region MOS transistor, the linearity performance will be degraded. In the equation, the first-order term defines the transconductance of the MOS resistor. The even-order harmonic terms can be cancelled out by the differential structure and thus the third-order harmonic distortion would become the dominant component. We can find that to minimize the third-order distortion term of the circuit, a small V_{cm} should be taken. In addition, the transistor M17 also works in the linear region to replace the resistor R_2 shown in Fig. 4.17.

Figure 4.19 shows a large signal simulation with respect to the function of the differential input voltage. The tuning range of 2–110 µS corresponding to the scaled value of current I_{tune} can be obtained. The tuning ratio of more than 55 can be achieved, and it is suitable for our applications. We can find that transistors M13 and M14 will operate in the weak inversion region to saturation region at large bias current, and the scaled function will become a radical expression.

The CMFB circuit, which is composed by transistors MF1 to MF8, is used for the differential structure. The feedback loop forces the output common-mode voltage to the desired value, and then the linearity of the input transistors in the following transconductor circuit will be maintained. We should note that the aspect ratio of the transistor M18 should be twice the value of the transistors MF1 and MF2.

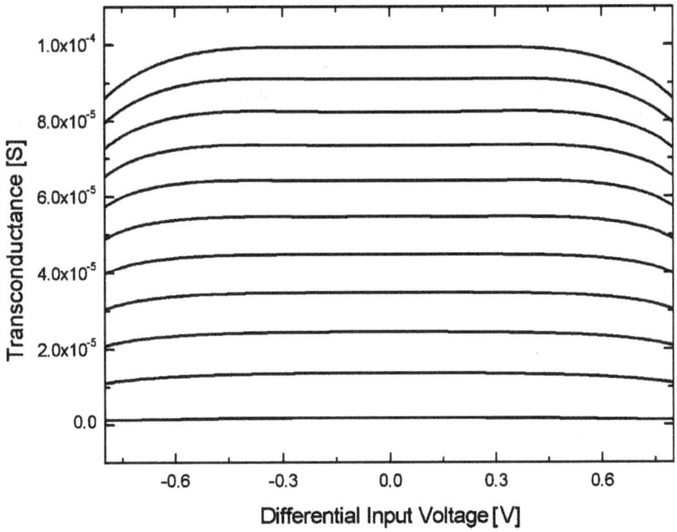

Fig. 4.19 The simulated Gm range of the proposed transconductor

4.5.2.4 Nonidealities in the Proposed Circuit

The thermal noise is a combination of the noise generated by the voltage-to-current conversion core and the current multiplier. In order to obtain the noise performance, a noise model validated in the weak and strong inversion region is required. As the model in [25] is introduced, from the analysis shows that the input-referred thermal noise contributed by the current multiplier can be reduced by a larger amplifier gain, where the amplifiers are composed by transistors M5 to M10 in Fig. 4.18. Therefore, the thermal noise can be reduced by increasing the transconductance of input transistors M5 to M8. Moreover, even the high linearity current multiplier contributes extra noise in the wide tuning transconductor, fortunately that half of the multiplier is combined with the voltage-to-current core, and thus only transistors M1 to M4 are added for current scaling. This condition reduces the contributed noise from the current multiplier.

The impact of flicker noise should also be taken into consideration. In this design, the device sizes of the transistors, which determine the noise performance, are increased to reduce the flicker noise, and the transistor length is designed to ten times of the minimal size used in the process. The larger size of the device also decreases the effect of channel length modulation and increases the output impedance.

The mismatch is the process that causes random variation in the physical quantities of identically designed devices. The random mismatch comes from the fabrication tolerances, and it produces the even-order distortion terms in differential structure. In the circuit, the linearity of the voltage-to-current conversion core is not sensitive to the mismatch owing to the feedback topology. The mismatch of the input transistor dimensions and threshold voltage will just produce a DC offset voltage at the resistor terminals. When the linear region transistors are introduced, the offset voltage will induce a slight variation of transconductance as the short channel effect is simply neglected, and then the linearity performance can be hold as well. In addition to the voltage-to-current conversion core, the mismatch problem will generate a higher nonlinear effect on the current multiplier. Simulation results show an increase of 3 dB in the second harmonic distortion term for a 2% mismatch of transistors M1 and M4. Thus, large transistor length is suitable for the multiplier in our design. This condition also helps the multiplier to achieve a higher linearity operation since it tends to work under weak inversion operation.

4.5.3 Filter Architecture and Measurement Results

From the demonstration of the passive ladder prototype, the third-order Butterworth low-pass G_m-C filter, which consists of seven identical transconductors, is used as shown in Fig. 4.5. The loading capacitor is realized from Metal-insulator-Metal structure and the unit of the capacitor is 1 pF. The −6 dB gain loss of the passive prototype has been compensated at the input of the filter by increasing the transconductance by a factor of 2.

The transconductor and the filter were designed in the TSMC 180 nm CMOS process. The chip micrograph is shown in Fig. 4.20 with the active area less than $0.23\,mm^2$.

The transconductor has been examined in the frequency domain to obtain the linearity performance. Figure 4.21 shows the spectrum of the transconductor through

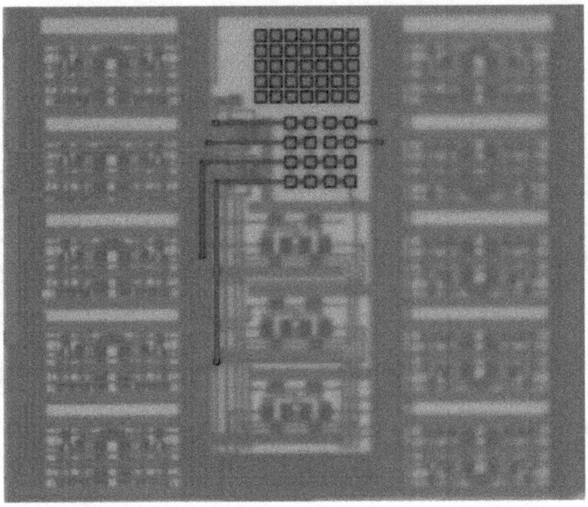

Fig. 4.20 The chip micrograph

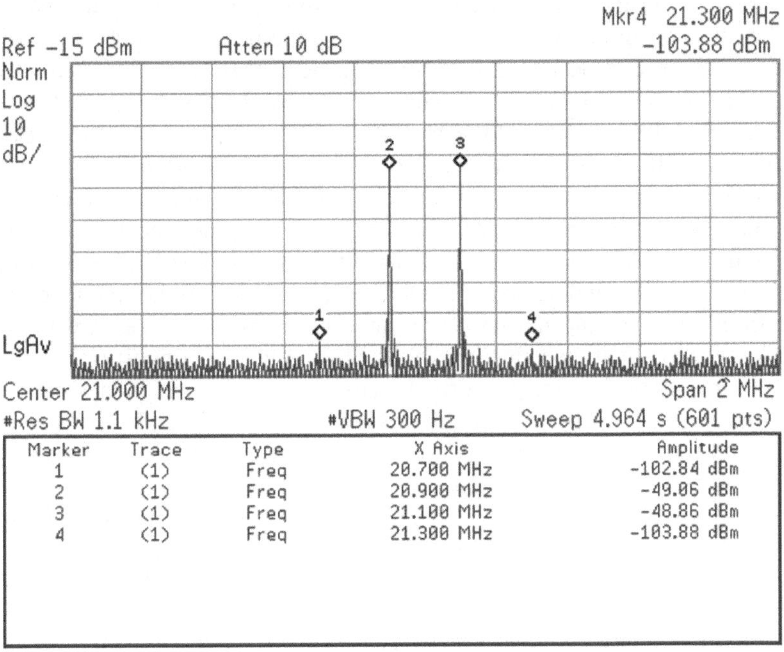

Fig. 4.21 Two-tone test of the proposed transconductor

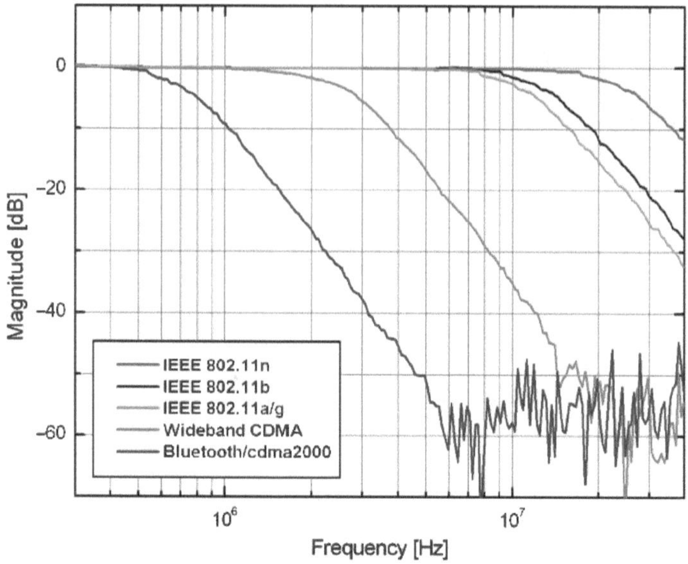

Fig. 4.22 The measured frequency responses of the proposed multi-mode filter

inter-modulation characterization by applying two tone signals near 20 MHz with the amplitude of 0.6 V_{pp} voltage. The result shows that the IM3 is around −54 dB.

Figure 4.22 illustrates the filter frequency response at 1.2-V supply. We should note that the magnitude is normalized owing to the use of the output buffer. The cutoff frequency can be tuned from 500 kHz to 20 MHz, and the range covers the specifications of Bluetooth, cdma2000, Wideband CDMA, and IEEE 802.11 a/b/g/n wireless LANs. The IM3 test yield the IP3 results of 22.3 dBm, 21.8 dBm, 20.5 dBm, 20.2 dBm, and 19 dBm for each wireless specification, respectively. The measured input-referred noise spectrum density at 70° C is summarized in Table 4.3. Since the cutoff frequency of the G_m-C filter is programmed by adjusting the transconductance, the filter can be called as the constant-capacitance network. The integrated output thermal noise is shown to be independent to the frequency scaling factor from analysis, and the dynamic range is almost the same over the tuning range. We can find that the value in Table 4.3 has an inverse proportion factor rather than a square-rooted inverse proportion factor to the scaled frequency. The deviation of the expected value is owing to the circuit flicker noise at low frequency and extra circuit on PCB board at high frequency. The measured noise gives an 80 dB dynamic range at IEEE 802.11 mode for −40 dB IM3. The DC offset decreases the dynamic signal swing, and then reduce the gain and linearity performance of the receiver. The offset cancellation loop/algorithm is usually used to compensate DC value while keeping the system performance well. In baseband filter design, DC offset also occurs owing to self correlation of input signals, and thus the second-order distortion performance would be important. The measured input second-order intercept point (IIP2), which can stand for the second-order distortion performance for

Table 4.3 Performance summary of this work

Technology	0.18-μm CMOS				
Supply	1.2-V				
Filter Prototype	3-rd order Buttorworth Lowpass Filter				
Tuning Range	500 kHz – 20 MHz				
Application	Bluetooth/ cdma2000	Wideband CDMA	IEEE 802.11 a/g	IEEE 802.11 b	IEEE 802.11 n
IIP3	22.3 dBm	21.8 dBm	20.5 dBm	20.2 dBm	19 dBm
IIP2	40 dBm	39.3 dBm	-	-	30.8 dBm
Out-of-band IIP3	17.5 dBm	17.3 dBm	-	-	13 dBm
Power	4.1 mW	4.7 mW	7.1 mW	8.2 mW	11.1 mW
Input-referred noise density	425nV/√Hz	128nV/√Hz	29nV/√Hz	24nV/√Hz	12nV/√Hz

Table 4.4 Comparison with previously reported works

Reference		[54] ESSCIRC 2003	[55] ESSCIRC 2004	[44] JSSC 2005	[56] JSSC 2006	This Work
Technology		0.18μ CMOS	0.18μ CMOS	0.25μ BiCMOS	0.13μ CMOS	0.18μ CMOS
Area		0.83mm²	0.125mm²	0.25 mm²	0.45mm²	0.23mm²
Type		G_m-C	G_m-C	G_m-C	Active-G_m-R C	G_m-C
Supply voltage		1.8	1.8	2.5	1.2	1.2
Filter order		6	4	3	4	3
-3dB frequency		1.5MHz-12MHz	0.5MHz-12MHz	50kHz-2.2MHz	2.11MHz-11 MHz	500kHz-20MHz
Tuning ratio		8	24	40	5.5	40
Application		W-CDMA/ IEEE 802.11a/b/g	BlueTooth /W-CDMA/ IEEE 802.11a/b/g/	GSM /BlueTooth /W-CDMA/ CDMA2000	UMTS/ IEEE 802.11a/b/g	BlueTooth/ cdma2000/W-CDMA /IEEE 802.11a/b/g/n
IIP3	**Min**	7.2dBm	9.4dBm	22dBm	21dBm	19 dBm
	Max	9.3dBm	11.1dBm	28dBm		22.3 dBm
Power	**Min**	10mW	1.1mW	2.5mW	3.4mW	4.1mW
	Max	15mW	4.5mW	7.3mW	14.2mW	11.1mW

every wireless mode, is also shown. Moreover, the linearity performance of out-of-band blocking interferences is measured. The value is described by out-of-band IIP3 in Table 4.3. In addition, Table 4.4 summarizes the filter type and detail information reported in recent year.

4.5.4 Summary

The CMOS implementation of a third-order Butterworth low-pass G_m-C filter for multi-mode applications is presented. The transconductor is designed by the combination of a voltage-to-current circuit and a current multiplier to achieve both the high linearity and wide tuning range simultaneously. Through the use of wide tuning range linear transconductor as a building block, the cutoff frequency of the channel selection filter can be widely tuned from 500 kHz to 20 MHz, which meets the specifications of Bluetooth, cdma2000, Wideband CDMA, and IEEE 802.11 a/b/g/n wireless LANs under direct-conversion architecture. The theoretical analysis of the high performance operation and a complete set of measurement results are provided to demonstrate the validity of the filter.

4.5.5 Analysis for Circuit at 1-V Supply

In this section, we discuss the performance when the supply voltage of proposed circuit operates from 1.2-V to 1-V. To make sure correct transistor operation region for reduced supply voltage, the input signal swing range should be defined. From Fig. 4.18, we can find that when small change of supply voltage is provided, both of the upper and lower boundaries have direct relationship to supply voltage, and thus the reduced supply voltage would not change the input swing range. Moreover, when the device is scaled down to 130 nm CMOS process, the upper boundary would be further relaxed by smaller threshold voltage. In addition, the transconductance of proposed circuit would become larger at lower supply voltage. The reason is the inverse relationship between the transconductance and input common-mode voltage as shown in (4.40) and (4.44). Under this condition, linearity of the proposed circuit can be increased from analysis shown in Section 4.5.2.3. Tuning can be achieved since the small change of supply voltage does not affect the function of translinear loop. Figure 4.23 shows the simulated Gm range of the proposed transconductor

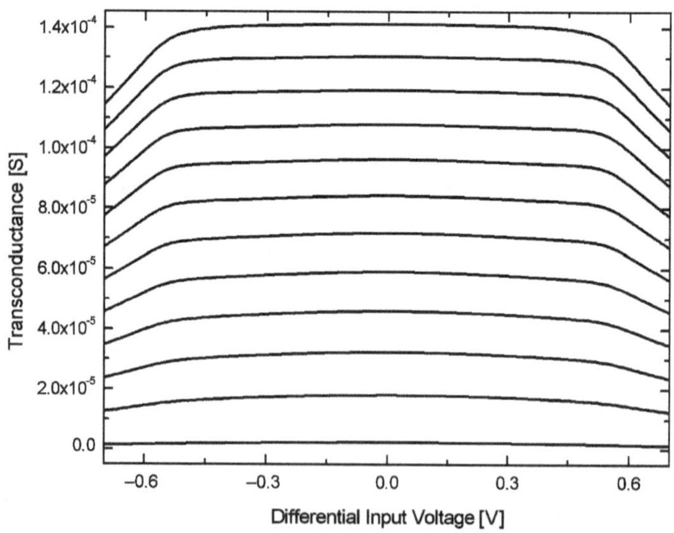

Fig. 4.23 The simulated Gm range of the proposed transconductor at 1-V supply

at 1-V supply. Compared with the 1.2 V swing range at 1.2-V supply in 180 nm CMOS process, the acceptable input swing range becomes to 1 V at 1-V supply in 130 nm CMOS process, and the same rail-to-rail operation is guaranteed. The large transconductance implies the implemented filter can work for higher cutoff frequency. From simulation analysis, the highest cutoff frequency can be extended to 40 MHz. The main disadvantage of the smaller supply voltage is the reduced gain performance. However, the CMRR can be still maintained since the tail current source, which is composed by transistor M18, is used at output stage. Simulation result shows the value of 67 dB is achieved.

Chapter 5
High Speed Filter with Automatic Tuning Circuit

5.1 Introduction

The high speed filters are discussed in the chapter. One application of the high speed filter is in the storage devices, such as the hard disk system. The signal from an HDD system can corrupt by distortion, noise and interference. The inter-symbol interference (ISI) would corrupt the signal source, and it would be the dominant non-ideality in high density detecting. The read channel, which performs the partial response maximum likelihood (PRML) method to reduce ISI, places an important role in the HDD system. Figure 5.1 shows the architecture of hard disk front-end. The preamplifier reduces the noise from the magnetic media, and then the signal is fed to the variable gain amplifier (VGA). The filter performs simple signal equalization. After simple equalization, an analog-to-digital converter (ADC) with 6-bit resolution is used. Finally, the digital signal processing (DSP) core performs complete signal equalization. The DSP core also controls the gain and timing through the interface. The modern approach uses a fifth-order filter for magnitude equalization, and other equalization is done in the digital processing. The fifth-order filter is composed by a one pole filter following the VGA and a fourth-order filter. The high speed fourth-order filter should have a linear phase. The details of a linear phase filter would be discussed in the next section. The other approach of high speed filter is the high speed wireless application, such as the ultra wideband (UWB) system.

The automatic tuning circuit, which compensates the effects caused by the corner variation in the fabrication process, is discussed in this chapter. Then, two implementations to the HDD and UWB system are introduced.

5.2 Linear Phase Filter

The purpose of the linear phase filter is to minimize the shifts in pulse peak for the binary data. In general, the linear phase under all signal frequencies maintains the data density. Non-uniform group delay would induce phase distortion and lead

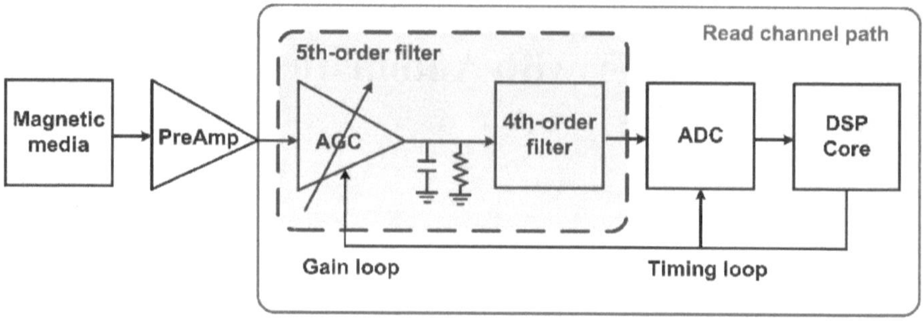

Fig. 5.1 Simplified architecture of HDD front-end

to detection problems. The other purpose is to reduce error rates by limiting the noise band from the detected media. The linear phase approximation can also be equivalent to the constant group delay. The ripple on the group delay is defined as the variation of the group delay relative to the normal group delay in a specified frequency band. In practice, the delay variation should be less than 5% over the filter passband and the order of 4–7 is mostly used.

5.2.1 Filter Transfer Function

Bessel-Thomson approximation and equiripple delay approximation are the two filter approximations for the filter with constant group delay. The equiripple delay approximation serves a 5% error until 1.5 times the cutoff frequency, rather than only the cutoff frequency of Bessel-Thomson approximation. For an equiripple delay filter, the delay error would be uniformly distributed over the frequency range, rather than decreasing it monotonically toward the corner of passband. Besides, the equiripple delay filter has better selectivity. The nominal transfer function of an equiripple delay filter is given by

$$H(s) = \frac{E(0)}{E(s)} = \frac{a_0}{s^n + a_{n-1}s^{n-1} + \cdots + a_1 s + a_0} \tag{5.1}$$

We should note that a closed form solution couldn't be found for a constant group delay. Since the phase of the transfer function will be linear and the sinusoidal ripple can be assumed. We can have

$$\theta(\omega) = -G_o\omega - \Delta\theta \sin(\omega T) \tag{5.2}$$

where G_o indicates a constant group delay value and T is the period of the ripple. The group delay is then given by

$$G(\omega) = \frac{-d\theta(\omega)}{d\omega} = G_o + \Delta\theta T \cos(\omega T) \tag{5.3}$$

Table 5.1 The denominator of equiripple linear phase transfer function

Filter order	E(s) of (5.1)
4	$(s^2+1.4894s+2.517)(s^2+1.9294s+1.1561)$
7	$(s+0.8613)(s^2+1.1456s+5.3703)(s^2+1.1542s+2.9514)(s^2+1.685s+1.3170)$

T is also equal to 2/n times the constant equal delay bandwidth, where n is the order of the filter. The delay error is given by

$$\Delta G = 2\Delta\theta T \tag{5.4}$$

It has twice the ripple width and is proportional to the phase error.

Table 5.1 shows the normalized transfer function for $\omega = 0.05°$. The pole locations are given, and the denormalization can be performed to reveal practical filter transfer function for a specified cutoff frequency. Since the value is obtained by the multiplication of a series of first order or second order function, the filter can be realized by cascading biquad sections.

For a high order equiripple filter, we can obtain a better group relay response, such as extended equiripple range in the frequency domain. However, there are some constraints when designing a high speed filter. The required high frequency poles in high order filter would be affected by parasitic poles. Thus, the delay error would largely reduce the expected transfer function.

5.2.2 *Group Delay Sensitivity*

In the section, the sensitivity of group delay variation in a biquad transfer function is discussed. The basic property of biquad filter is discussed in Chapter 2 and the transfer function is given by

$$H(s) = \frac{\omega_o^2}{s^2 + s\frac{\omega_o}{Q} + \omega_o^2} \tag{5.5}$$

We let s = $j\omega$ for real frequencies, H(s) is given by

$$H(j\omega) = M(\omega)\, e^{j\Phi(\omega)} \tag{5.6}$$

The gain sensitivities of the biquad to ω_o and Q are [57]

$$S_{Q}^{M(\omega)} = \left[1 + Q^2\left(\frac{\omega}{\omega_o} - \frac{\omega_o}{\omega}\right)^2\right]^{-1} \tag{5.7}$$

$$S_{\omega_o}^{M(\omega)} = S_{Q}^{M(\omega)}\left[1 + 2Q^2\left(\frac{\omega^2}{\omega_o^2} - 1\right)\right] \tag{5.8}$$

Besides the group delay sensitivities to ω_o and Q can be expressed as [58]

$$S_{Q}^{G(\omega)} = -1 + 2S_{Q}^{M(\omega)} \tag{5.9}$$

$$S_{\omega_o}^{G(\omega)} = -3 + 2S_{\omega_o}^{M(\omega)} + \frac{2}{1 + \frac{\omega^2}{\omega_o}} \tag{5.10}$$

For an nth order low-pass biquads, the group delay variation can be given by

$$\frac{\Delta G(\omega)}{G(\omega)} = \sum_{i=1}^{N}\left(S_{Q_i}^{G(\omega)_i}\frac{\Delta Q_i}{Q_i} + S_{\omega_{o,i}}^{G(\omega)_i}\frac{\Delta\omega_{o,i}}{\omega_{o,i}}\right)\frac{G(\omega)_i}{G(\omega)} \tag{5.11}$$

If we use the fourth order equiripple filter shown in Table 5.1 as an example, we can obtain the group delay sensitivity for each of the biquad from (5.9) and (5.10). The sensitivities to both the Q_1 and ω_{o1} are less than 1 for the first biquad, and the sensitivity of Q_2 are less than 1 for the second biquad. However, the sensitivity of ω_{o2} are larger than 1 at higher frequency. In the G_m-C implementation, the Q value can be designed by the aspect ratio of the transconductance and it is nearly independent to the process. The resonant frequency ω_o will vary under corner variation, and thus an automatic frequency tuning circuit can be used to reduce the group delay variations.

5.3 Automatic Tuning Circuit

Automatic tuning circuit is an important component for continuous-time filters. Usually, the RC time constant variation of more than 30% would be provided. Thus, the tuning circuit should be used to compensate for the deviation from process and temperature that affect filter accuracy. The frequency tuning is the most important approach in the automatic tuning circuit since the filter cutoff frequency determinates the system performance. However, the filter quality factor determines the gain of the filter, and the Q tuning would be sometimes important especially in the narrow band applications. For the G_m-C implementation, we need the tuning circuit to modify the property of transconductance or loading capacitance. We should note that although the individual components are largely affected by the process variation, the ratio of the same components is still matched.

In the automatic tuning scheme, the filter characteristic is measured at first, and it will be compared with the desired value. Then, the corrective feedback signal is

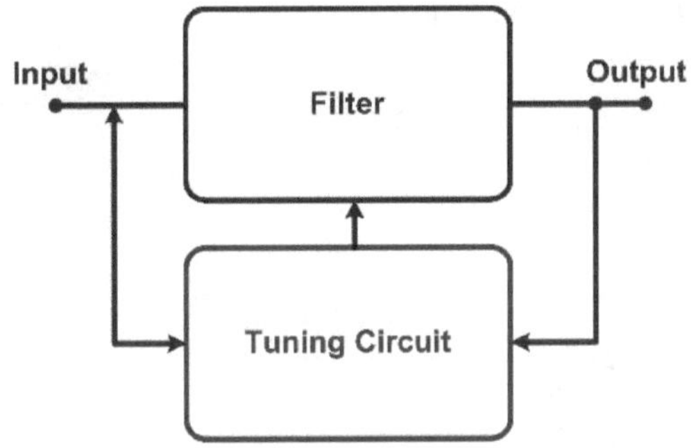

Fig. 5.2 Direct tuning scheme

applied to reduce the error to zero. There are two kinds of tuning architectures. One is the direct tuning architecture and the other is the indirect tuning architecture.

5.3.1 Direct Tuning Architecture

Figure 5.2 shows the direct tuning architecture. The concept of direct tuning is to use the filter which processes the actual signal. If continuous operation is required, the tuning algorithm should be able to avoid the effect of the filter transfer characteristic. This architecture is similar to the adaptive tuning techniques. However, if the filter operates under the sleep mode and be removed from the signal path at some times, the tuning can become simpler. This kind of tuning technique can be employed in video filters, where tuning can be performed in the field fly-back interval [59].

5.3.2 Indirect Tuning Architecture

Since the tuning algorithm is very complicated, the indirect tuning architecture becomes popular and it is commonly used in circuit implementation. The indirect tuning can be referred to as the master-slave tuning. In other words, the filter is not directly tuned by using the filter output signal. The indirect tuning architecture is shown in Fig. 5.3. The slave filter that processes signal is left alone. The master block could be a transcontuctor or a filter. Any non-idealities of the master block would produce the same effects in the slave filter. If the characteristics of master block are corrected, the same information would be applied to the slave filter.

The first approach based on the integrator is discussed. Figure 5.4a shows the constant transconductance tuning. In the approach, the transconductance is set to the inverse of an external resistance. The circuit works as follows: if G_m of the transconductor is small, the current through R_{ext} is larger than the current supplied by the transconductor, and the difference between the two currents is integrated with the OPAMP and capacitor. Finally, the control voltage, V_{ctrl}, is increased until the

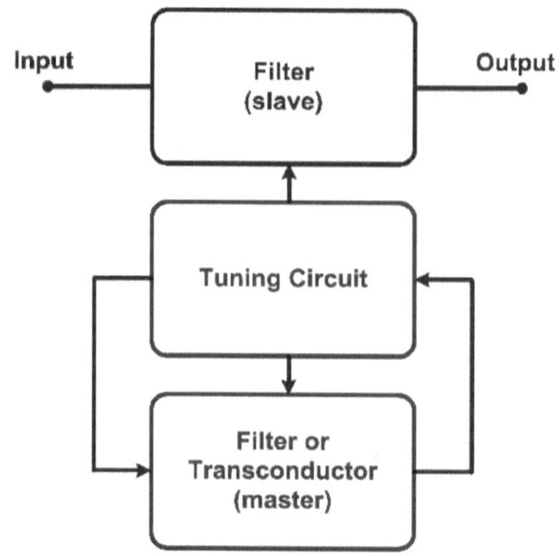

Fig. 5.3 Indirect tuning scheme

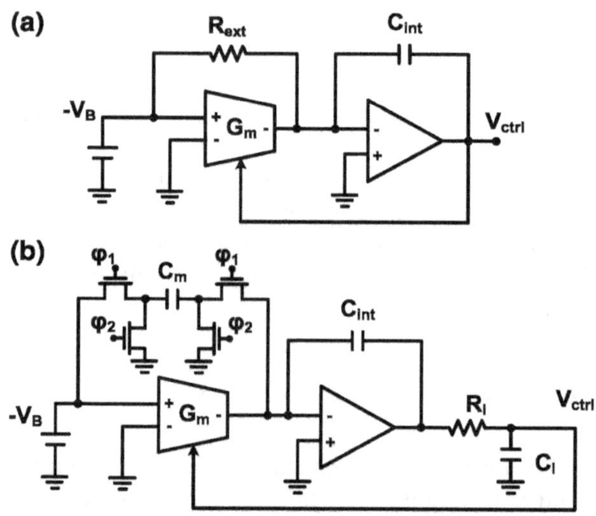

Fig. 5.4 Frequency tuning circuit by using single transconductor: **(a)** resistor based tuning **(b)** switch based tuning

same transconductance is obtained. Instead of the passive resistor, we can use the switch-capacitor circuit. Figure 5.4b shows the modified circuit, and the equivalent resistance in Fig. 5.4 is given by $R_{ext} = 1/(f_{clk}C_m)$. Thus, the transconductance is set to $f_{clk}C_m$. Under this condition, a precise tuning circuit can be achieved since G_m/C_L is set to $f_{clk}C_m/C_L$, and then the ratio of the capacitance can be hold. We should note that an additional low-pass filer has been added to remove the high frequency ripple voltage owing to the switched technology. However, some clock jitter would still leak into the slave filter through the control node. The other

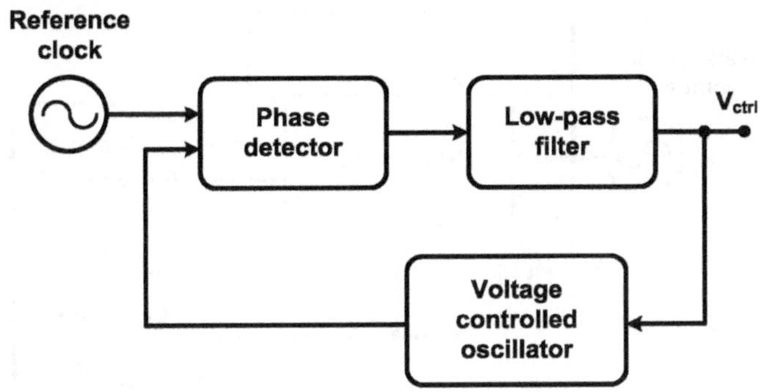

Fig. 5.5 Frequency tuning based on the VCO

disadvantages of the approach for high speed filter are the required large capacitor ratio, which implies poor matching problem, and the high speed clock reference.

Another approach is based on the voltage controlled oscillator (VCO) technique. The tuning scheme is shown in Fig. 5.5. In the technique, two integrators are placed in a loop. Then, a phase lock loop (PLL) is used to achieve equal signal frequency between the external reference and oscillator. In other words, the oscillator frequency of the VCO is equal to the external reference, and so does the filter pole frequency through the corrected voltage, V_{ctrl}. The phase detector can be a simple XOR gate since only the phase variation should be detected. The problem with the VCO technique is the limited oscillation amplitude. Usually, a second-order harmonic oscillator is used, and the harmonic distortion of transconductor will shift the effective oscillation frequency. Thus, an amplitude regulation should be provided to make sure the high linearity operation of the transconductor. For best matching between the tuning circuitry and the slave filter, we should choose equal values for reference signal frequency and cutoff frequency. However, the noise typically provides the largest value at filter cutoff frequency, and therefore the signal leakage would largely degrade the filter dynamic range.

The voltage controlled filter (VCF) approach is another choice. The master block of the approach is composed by a second-order biquad low-pass filter, as shown in Fig. 5.6. Since a phase shift of 90° can be obtained for the biquad, the multiplication of the phase should be equal to zero at the locked condition. It means that the frequency dependent input to output phase characteristics of the reference are exploited to tune the circuit. For this reason, any offset in the phase comparator will result in a frequency tuning error. The performance, such as accuracy and speed, of the comparator should be maintained. In this approach, a high performance reference signal should also be required. The reason is that the harmonics will not behave the same phase shift as the fundamental frequency, and can thus corrupt the result of phase comparison.

The above frequency tuning can be applied by tuning the transconductors continuously. The transconductor should be designed to have specified linearity over the tuning range, and much effort of the transconductor circuit should be taken.

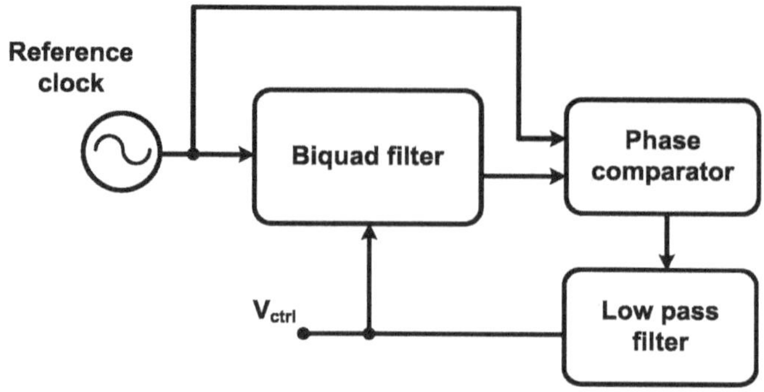

Fig. 5.6 Frequency tuning based on the VCF

In addition to the tuning scheme performed in the analog domain, the digital frequency tuning scheme could be another issue. Through the use of programmable transistor or capacitor arrays, the linearity can be easily maintained at the expense of additional area and accuracy. In the programmable filter, the analog tuning circuit requires additional interface since the output is a voltage. Thus, the digital tuning circuit, which is composed of the counters and control logics, can be adopted for the programmable filter to control switches.

5.4 A 1 GHZ Equiripple Low-Pass Filter with a High-Speed Automatic Tuning Scheme

A continuous-time fourth-order equiripple linear phase G_m-C filter with an automatic tuning circuit is presented. A high speed transconductor based on the inverter structure is realized. The combined CMFF and CMFB circuit ensures the input and output common-mode stability. The gain performance could be maintained by combining a negative resistor at the output nodes. Transconductance tuning can be achieved by adjusting the bulk voltage by using the Deep-NWELL technology. Through the use of the transconductor as a building block with a modified automatic tuning scheme, the filter −3 dB cutoff frequency is 1 GHz with the group delay less than 4% variation up to 1.5 fc. The −43 dB of IM3 at filter cutoff frequency is obtained with −4 dbm two tone signals. Implemented in 180 nm CMOS process, the chip occupies an area of 1 mm^2 and consumes 175 mW at a 1.5-V supply.

5.4.1 Introduction

The continuous-time filters have been widely used in various high speed applications, such as high data-rate read channel hard disks, wireline and wireless communications. The −3 dB cutoff frequency of the designated low-pass filter needs to increase with the speed of the applications, but the high performance constraints, such as low power dissipation, small area, high linearity, and flat group delay, are not at all lost. The G_m-C topology with simplicity, modularity, open loop configuration, and electronic tunability would be the conspicuous choice for high frequency filter design.

The transconductor is the main building block in the G_m-C filter topology. The key function of the transconductor is to convert the input voltage into the output current while accuracy and linearity are both maintained. However, the non-idealities of transconductor dominate the filter performance. The parasitic capacitors produce the deviation of −3 dB cutoff frequency of low-pass filters, the finite output impedance affects the quality factor, and the voltage-to-current conversion affects the filter linearity. Many previous works have been proposed, but those highly linear circuits are difficult to be used when high speed is required.

In this section, a 1 GHz fourth-order equiripple linear-phase G_m-C low-pass filter, based on a high performance transconductor, with an automatic tuning circuit is presented. Section 5.4.2 develops the high speed transconductor based on the Nauta's inverter structure [60]. Owing to the pseudo-differential structure, the suitable common-mode control system should be included. In Section 5.4.3, the proposed filter is designed by cascading two biquadratic filters. A novel automatic tuning circuit designed to suppress the effects caused by the fabricated corner variation and temperature is also discussed in the same section. The measurement results are shown in Section 5.4.4. Finally, the results are summarized in Section 5.4.5.

5.4.2 Operational Transconductance Amplifier

5.4.2.1 The Voltage-To-Current Conversion

The class-AB transconductor circuit is shown in Fig. 5.7. The voltage-to-current conversion circuit is composed of transistors M1 to M4, where the device parameter

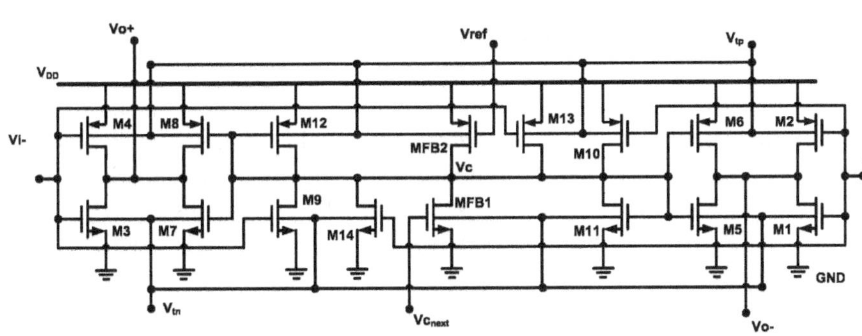

Fig. 5.7 The high speed transconductor circuit

of M1 is equal to M3 and M2 is equal to M4. These transistors would operate in the saturation region, and the signals applied to the gate terminal are $V_{i+} = V_{cm} + v_d/2$, and $V_{i-} = V_{cm} - v_d/2$ (V_{cm} is the input common-mode voltage and v_d is the input differential voltage). By using the square law equation of saturated transistors, the voltage-to-current conversion can be analyzed and the transconductance can be expressed as

$$G_m = \left(V_{DD} - V_{thn} + V_{thp}\right)\sqrt{K_{1,3}K_{2,4}} \tag{5.12}$$

where V_{DD} is the supply voltage, V_{thn} is the threshold voltage of the NMOS transistor, V_{thp} is the threshold voltage of the PMOS transistor, and K_i is the device parameter of transistor M_i. Thus, the transconductance will be related to device parameter, supply voltage, and threshold voltage. We should note that when large device sizes are used, the transconductor is linear even though $K_{1,3}$ is not equal to $K_{2,4}$.

For the circuit to operate at very high frequency, the absence of internal nodes would be significant to avoid the effect caused by the parasitic capacitance. We can find that the simple voltage-to-current conversion composed of transistors M1 to M4 operates with no internal nodes. Thus, the only parasitic capacitance existing in the signal path is due to the transistor channel, and the pole would locate at the tens of GHz range. In addition, a large transconductance should be designed because the transconductance would proportional to the −3 dB cutoff frequency of the G_m-C low-pass topology. With small feature sizes of nano-scale CMOS technology, the drain current of single MOS transistor can be approximated as

$$I_{D,sat} = \frac{K\left(V_{GS} - V_{th}\right)^2}{2\left[1 + \theta\left(V_{GS} - V_{th}\right)\right]} \tag{5.13}$$

where θ is the mobility reduction coefficient. If we assume that $K_{1,3}$ is equal to $K_{2,4}$, we can define $V_{ov} = V_{cm} - V_{thn} = V_{DD} - V_{cm} - |V_{thp}|$. From the analysis of a Taylor series expansion, the third-order harmonic distortion term would be the dominant component of transconductor, and the HD3 is given by

$$HD_3 = \frac{-2\theta v_d^2}{V_{ov}\left[4 + 2\theta V_{ov}\left(5 + 4\theta V_{ov} + \theta^2 V_{ov}^2\right)\right]} \tag{5.14}$$

Thus, the linearity can be improved by giving a larger overdrive voltage, which requires a higher supply voltage or a smaller threshold voltage. Usually, half the value of supply voltage should be chosen for input common-mode voltage when $K_{1,3}$ is close to $K_{2,4}$. In the situation, the output voltage would be close to half of the supply voltage, and thus no output common-mode current would appear.

For the required large transconductance, the thermal noise, which dominates noise performance for high speed circuit, would be reduced as well.

5.4.2.2 The Common-Mode Control System

The transconductor behaves as a pseudo-differential structure and thus a CMFF circuit should be used to restrict the effect caused by the variation in the input common-mode signal. Transistors M9 and M14 have one half of the device parameter of transistor M1, and transistors M10 and M13 have one half of the device parameter of transistor M2, respectively. The input common-mode signal can be obtained by using transistors M9, M10, M13 and M14. Then, the quantity of the input common-mode variation would be cancelled out at the output nodes through transistors M11 and M12.

Owing to the cascading structure of G_m-C topology, the output nodes of one transconductor would be the input nodes of the following transconductor. The output common-mode voltage should be fixed to the value of the input common-mode voltage, and thus the linearity of the designed filter will be held. In our circuit, the output common-mode voltage is maintained by an adaptive CMFB circuit, which includes transistors MFB1 and MFB2. No additional common-mode sensing circuit would be required because the information regarding output common-mode signal would appear at the next transconductor stage. Thus, the output common-mode information will be detected by the voltage, Vc_{next}, which is the Vc node of the next transconductor stage. Then, the signal produced by the CMFB circuit would be combined with the CMFF circuit to adjust the output common-mode voltage accordingly.

For the open-loop simulation of common-mode system, a large inductor is placed in the loop in order to open small signals. In the system, Vc node introduces the non-dominate pole, and a phase margin of 53° can be obtained while the loading capacitor of 0.9 pF is given.

5.4.2.3 Gain Enhancement and Transconductance Tuning Circuit

In the topology, the gain performance should also be taken into account. The gain enhancement stage is restricted because it does not require any internal nodes in the high speed design. Moreover, the small feature size, which is chosen for small parasitic capacitance, also degrades the gain performance. The negative resistance circuit for gain enhancement, composed by transistors M15 to M18, is shown in Fig. 5.8a. With the addition of Fig. 5.8a, the fabricated DC gain of larger than 35 dB could be achieved. Besides, the channel length modulation effect, which is a distortion component contributed to the proposed circuit, can be minimized.

Without the existence of internal nodes, the possible transconductance tuning nodes left are the supply voltage and bulks. In [60], the transconductance was tuned by adjusting the supply voltage. However, the method not only degrades the linearity when a fixed common-mode voltage is applied from the previous stage of the system, but also increases the complexity of the regulator due to class-AB operation. Thus, the transconductance could be tuned by adjusting both the bulk voltage of PMOS and NMOS in the Deep-NWELL CMOS process. Figure 5.8b shows the

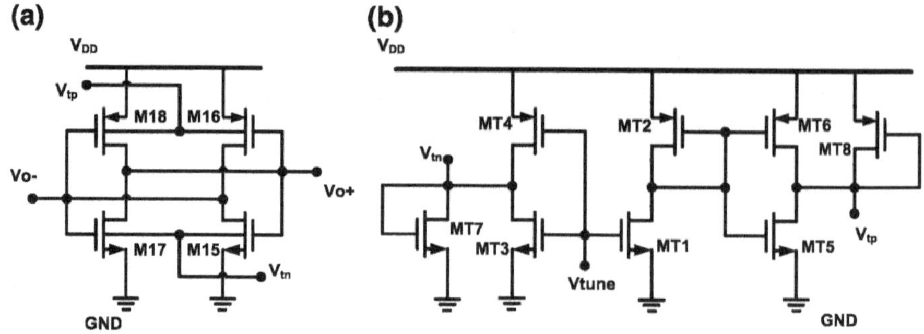

Fig. 5.8 Circuit implementation: **(a)** negative output resistance circuit for gain enhancement **(b)** transconductance tuning circuit

bulk tuning circuitry. When the voltage at V_{tune} is changed, the voltage at nodes V_{tn} and V_{tp} would be adjusted to the opposite values accordingly. The forward bias scheme would increase the value of V_{thn} and decrease the value of V_{thp}, and then the transconductance of the proposed transconductor would be dependent on the value of V_{tune}. There are some advantages of forward bias scheme. Firstly, the speed can be enhanced to a higher value while the increased power consumption is less than increasing the V_{DD} voltage. Secondly, the variation of threshold voltage becomes smaller because forward bias shrinks the depletion layer of MOS transistors [61]. Therefore, the short channel effect can be reduced. Finally, the overdrive voltage becomes large under this condition, and the linearity of the transconductor could be further increased.

However, the latch-up effect and leakage current may create problems, and thus the constraint of a 0.5 V forward bias in deep N-WELL process needs to be maintained [62]. Thus, we can design the device aspect ratio of transistors MT7 and MT8 by the following constraint

$$\left(\frac{W}{L}\right)_{MT7} \geq \frac{\mu_p}{\mu_n}\left(\frac{0.5 - V_{thn}}{V_{DD} - |V_{thp}|}\right)^2 \left(\frac{W}{L}\right)_{MT4} \tag{5.15}$$

$$\left(\frac{W}{L}\right)_{MT8} \geq \frac{\mu_n}{\mu_p}\left(\frac{0.5 - |V_{thp}|}{V_{DD} - V_{thn}}\right)^2 \left(\frac{W}{L}\right)_{MT5} \tag{5.16}$$

where μ_n and μ_p are the low-field mobilities of NMOS and PMOS transistors. We should note that the transistors MT7 and MT8 would operate in the weak inversion region in the circuit when a smaller forward bias voltage is applied. In addition, the value of the negative resistance for gain enhancement could be tuned separately by applying another bulk tuning circuitry, and thus the Q tuning can also be achieved.

5.4.3 Filter Architecture and Automatic Tuning Circuit

The architecture of the fourth-order equiripple linear phase filter is shown in Fig. 5.9. The filter is based on two-cascade of two-biquadratic filters. The LC ladder structure is chosen due to its low sensitivity. A constant group delay should be maintained to avoid detection problems in the frequency band where the spectral components of the signal are located. In the structure, each transconductor has its individual CMFF circuit. The number of CMFB circuits is reduced due to the sharing of the same output nodes. In order to obtain the stability performance of the biquad section, a current pulse is given at the output nodes and we can find that the 1% settling-time is less than 0.5 ns. The parasitic capacitors result in the deviation of cutoff frequency and the effect becomes prominent especially for the target of GHz application. Therefore, the integration capacitance must be designed by taking the transistor gate capacitance, junction capacitance, and additional MIM capacitance into consideration.

Since the transconductance and the capacitance change with process and temperature variations, an automatic tuning scheme should be used to maintain the time constant in this low Q filter design. The indirect tuning, which takes the advantage

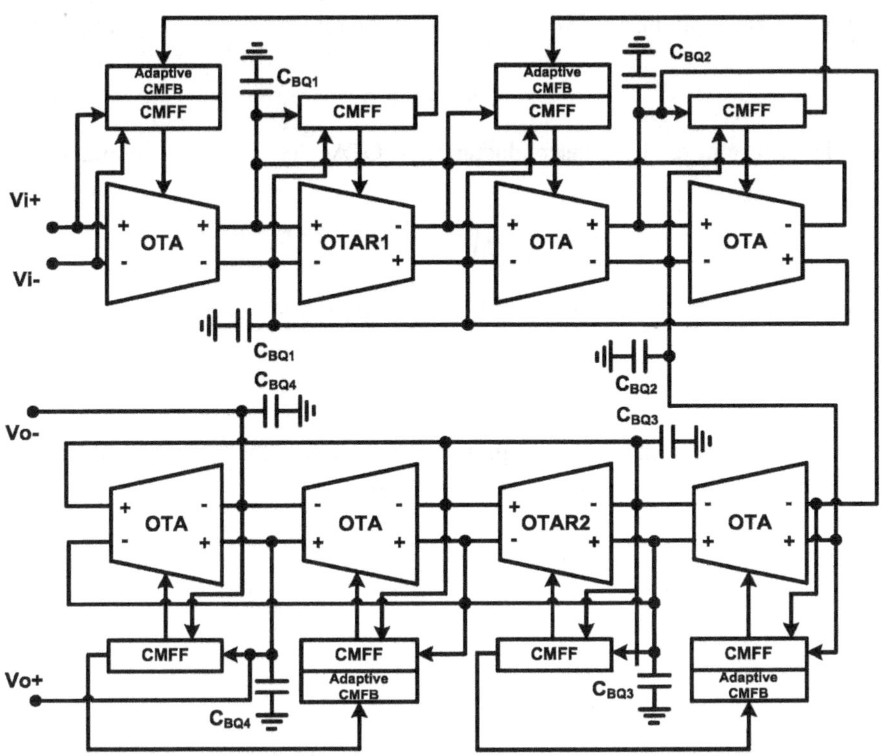

Fig. 5.9 The fourth-order equiripple linear phase filter

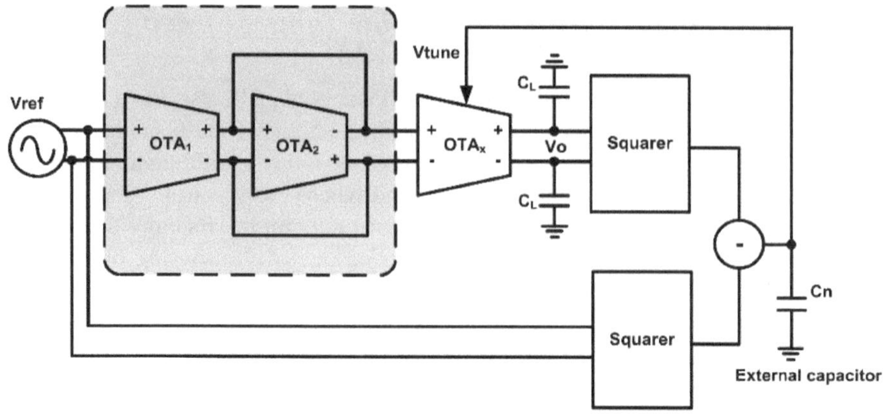

Fig. 5.10 The modified automatic tuning scheme

of less complexity and smaller area than the direct tuning, is used here. Figure 5.10 shows the proposed master-slave tuning strategy. The tuning strategy is composed by the transconductors, squarers, and a comparator. In the figure, OTA_x is a replica of the transconductor in the proposed filter and the same load condition of C_L should be applied. By applying a reference signal with the reference frequency of f_{ref}, which can be given by

$$v_{ref} = A \sin\left(2\pi f_{ref} t\right) \tag{5.17}$$

and defining gmi as the transconductance of OTA_i, the integrator output voltage becomes

$$v_o = \left(\frac{gm1}{gm2}\right)\left(\frac{f_u}{f_{ref}}\right) A \cos\left(2\pi f_{ref} t\right) \tag{5.18}$$

where f_u is the unity-gain frequency of the integrator and is given by

$$f_u = \frac{gmx}{2\pi C_L} \tag{5.19}$$

Then, the scheme utilizing the magnitude detection and the error current signal is generated based on the difference of the magnitudes

$$v_o^2 = \frac{1}{2}\left[A\left(\frac{gm1}{gm2}\right)\left(\frac{f_u}{f_{ref}}\right)\right]^2 + \frac{1}{2}\left[A\left(\frac{gm1}{gm2}\right)\left(\frac{f_u}{f_{ref}}\right)\right]^2 \cos(4\pi f_{ref} t) \tag{5.20}$$

$$v_{ref}^2 = \frac{1}{2}A^2 - \frac{1}{2}A^2 \cos\left(4\pi f_{ref} t\right) \tag{5.21}$$

Finally, a following low-pass filter would be used to filter out the high frequency components. When gm1 and gm2 have the transconductance ratio of k, the frequency of reference signal can be relaxed by the same ratio. Therefore, we can make the selection of the reference frequency flexible, rather than the only choice of the cutoff frequency in [63]. Besides, low frequency reference signal can be used to relax the high speed requirement of the tuning circuitry for our high speed filter, and the process corner variation would not affect the ratio of k largely. Finally, the control signal for correct transconductance is also applied to the slave integrator which matches the proposed master filter.

5.4.4 Measurement Results

The filter was fabricated by the TSMC 180 nm CMOS process and measured at a 1.5-V supply. The on-chip input and output buffers are used for the high frequency measurement. Figure 5.11 shows the measured magnitude response of the proposed filter. The cutoff frequency is set to 1 GHz by the automatic tuning circuit. Figure 5.12 shows the measured group delay characteristics. It shows that the deviation of the group delay is within the range of ±200 ps up to 1.5 fc. Figure 5.13 shows the −43 dB of IM3 at filter cutoff frequency with −4 dBm two tone signals of 0.995 and 1.005 GHz. The measured CMRR of the filter is 32 dB, and the dynamic range

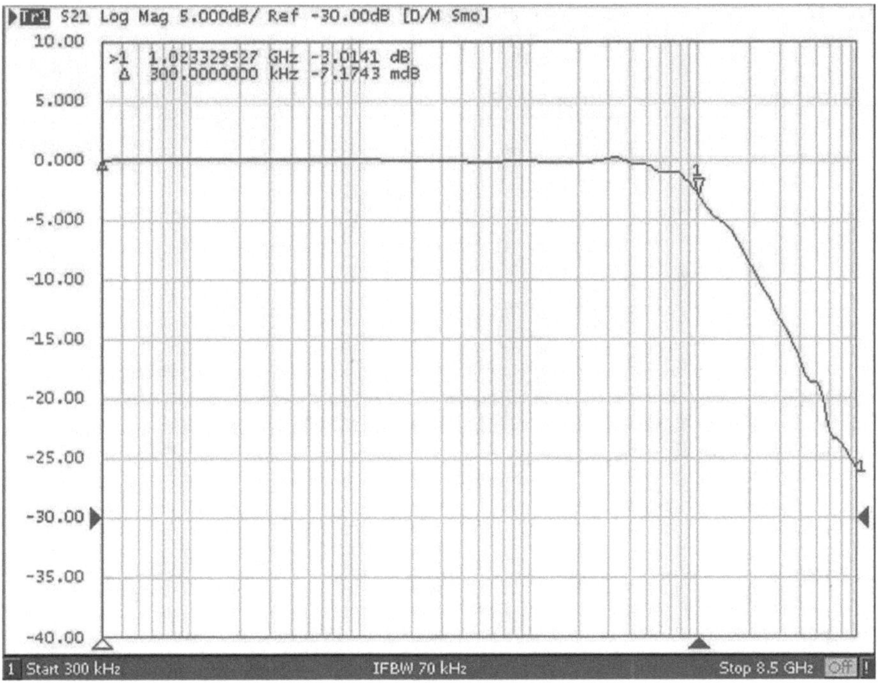

Fig. 5.11 The measured frequency response of the proposed filter

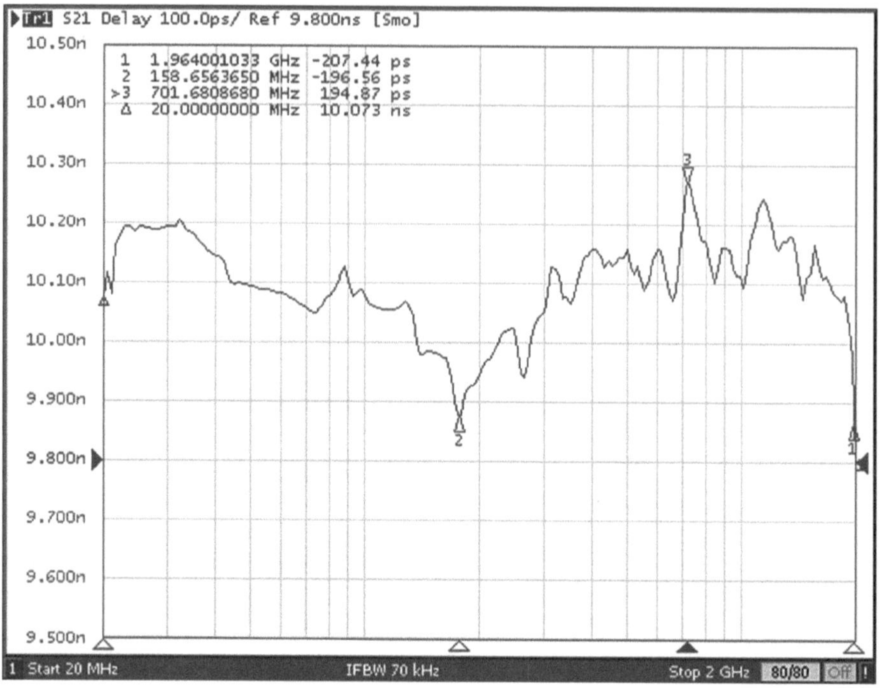

Fig. 5.12 The measured group delay of the proposed filter

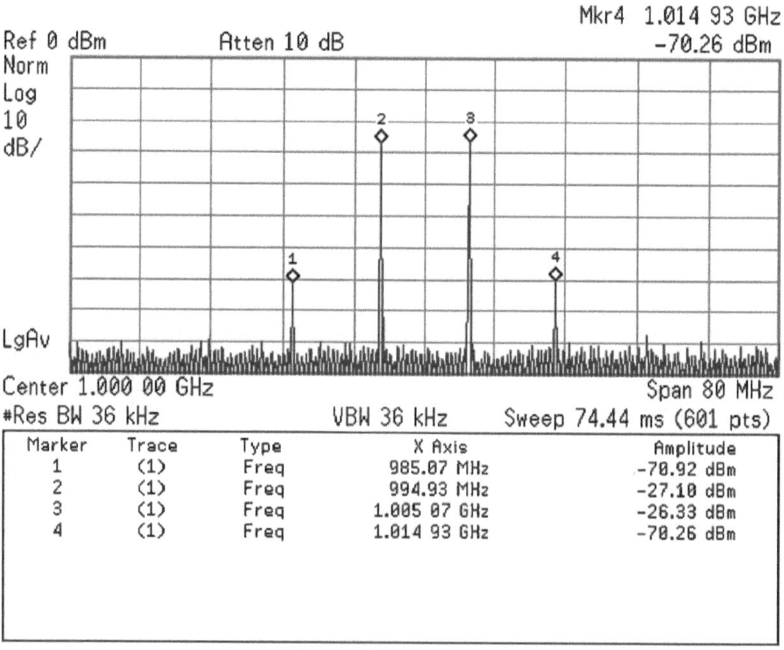

Marker	Trace	Type	X Axis	Amplitude
1	(1)	Freq	985.07 MHz	-70.92 dBm
2	(1)	Freq	994.93 MHz	-27.10 dBm
3	(1)	Freq	1.005 07 GHz	-26.33 dBm
4	(1)	Freq	1.014 93 GHz	-70.26 dBm

Fig. 5.13 Two tone inter-modulation of the filter

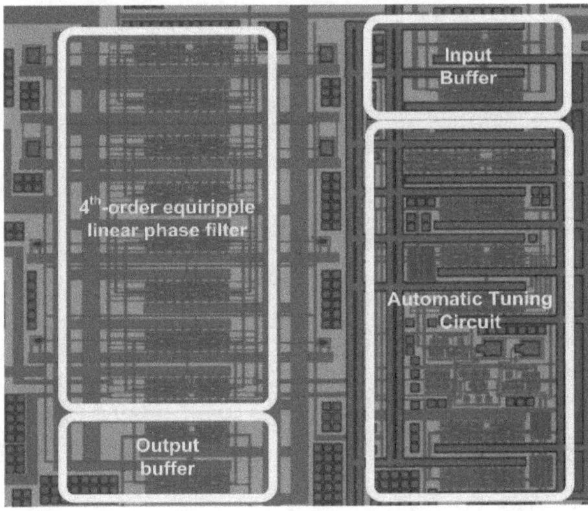

Fig. 5.14 Die micrograph

of 39 dB is measured at −43 dB IM3 performance. The filter and the automatic tuning circuit together dissipate 175 mW. The chip micrograph with the active area of 1 mm^2 is shown in Fig. 5.14.

To compare the proposed filter with those reported previously, the FOM defined in [64] is introduced. The FOM, which takes the boosting factor, the speed of filter, technology feature size and power per pole quantity into consideration, is given by

$$FoM = \frac{B\,[Bandwidth(MHz)]^2 \times technology(\mu m)}{ppp(mW)} \tag{5.22}$$

where B is the boosting factor. The boosting factor is assumed to be 1 for no-boosting structure and 1.5 if the reported filter has boosting. The filter results are compared with the previous realization in Table 5.2, and FOM shows the high performance operation of the proposed filter.

5.4.5 Summary

A high speed transconductor based on the inverter structure is realized. The combined CMFF and CMFB circuit ensures the input/output common-mode stability. The gain performance could be maintained by combining a negative resistor at the output nodes. Transconductance tuning can be achieved by adjusting the bulk voltage by using Deep-NWELL technology. The transconductor is used to design a 1 GHz fourth-order equiripple linear phase G_m-C filter. A modified automatic

Table 5.2 Comparison of previously reported works

Reference	2003 JSSC [63]	2006 TCAS-I [64]	2006 ESSCIRC [65]	This work
Technology	0.35-μm CMOS	0.35-μm CMOS	0.25-μm CMOS	0.18-μm CMOS
Filter order	7	4	7	4
-3dB frequency	200MHz	550MHz	200MHz	1GHz
Automatic tuning (AT)	Yes	Yes	Yes	Yes
Group delay ripple	<5% at 1.5fc	N/A	<5% at 1.5fc	<4% at 1.5fc
IM3/HD3	-44dB at 0.5Vpp	-40dB at 0.3Vpp	-42dB at 0.8Vpp	-43dB at 0.35Vpp
Supply	3V	3.3V	3.0V	1.5V
Power consumption	60mW	140mW	270mW	175mW
FOM	1633	3025	500	4114

tuning circuit which relaxes the need of high speed operation of the squarers and comparators is introduced. The theoretical properties of the proposed filter are experimentally verified.

5.4.6 *Analysis for Circuit at 1-V Supply*

When the supply voltage is reduced from 1.5-V to 1-V, the transistors used in the transconductor will operate from saturation region to weak inversion region. There are two problems when this circuit operates in the weak inversion region. One is the reduced transconductance. The reduced transconductance will lead to smaller unity-gain frequency and then the filter cutoff frequency. The other is the linearity performance since the exponential V-I relationship of the weak inversion devices will lead to larger distortion. If we increase the transistor length, the threshold voltage would be reduced, and the transconductor can work under the saturation region. At this condition, the linearity performance of the transconductor will be better than in the weak inversion region owing to squared V-I relationship. However, compared with the original circuit in 1.5-V supply, the overdrive voltage is smaller and thus the linearity caused by short channel effect will be worse. Also, the transconductance will further be decreased as derived from (5.12).

When the device size is scaled down to 130 nm CMOS, the decreased transconductance at 1-V supply can be recovered owing to smaller threshold voltage. Therefore, the transistors will work under saturation region. Although short channel effect affects the linearity, the overdrive voltage is larger than the same supply in 180 nm CMOS. Figure 5.15 shows the simulation result when the transconductor operate at 1-V supply. In 180 nm CMOS process, the transconductor under saturation region has smaller transconductance but better linearity than the transconductor

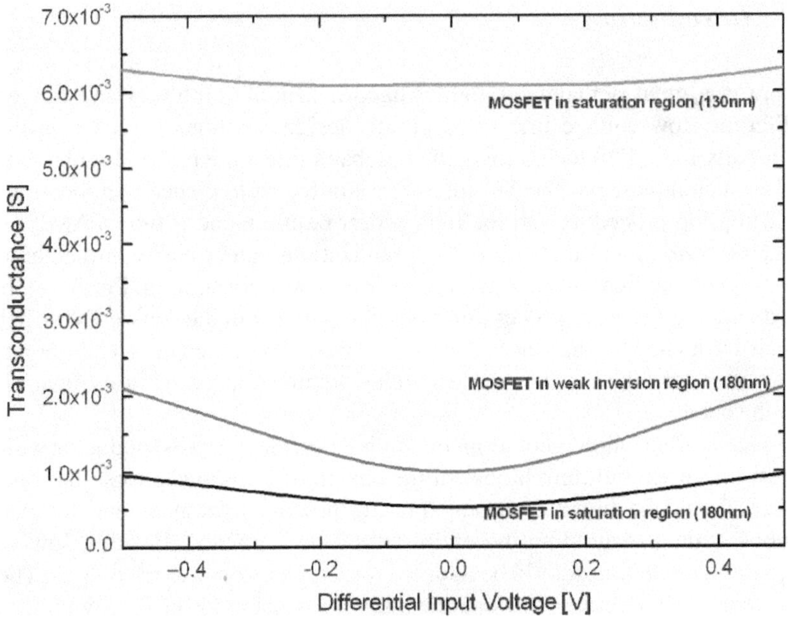

Fig. 5.15 The transconductance of the proposed transconductor at 1-V supply

under weak inversion region. On the contrary, the transconductor in 130 nm CMOS process behaves better transconductance and linearity. Simulation results also show the unity-gain frequency is 1.6 GHz and a small excess phase of 0.04 degree. The CMRR is 44 dB at low frequency. In addition, tuning can still be achieved in 130 nm CMOS process, but a limited range due to smaller threshold voltage would be provided.

5.5 A G_m-C Low-Pass Filter for UWB Wireless Application

This section proposes a high performance G_m-C equiripple linear phase low-pass filter for UWB wireless application. The proposed transconductor is designed under low supply voltage consideration while its gain, excess phase, and linearity are well maintained. The common-mode control system, including common-mode feedback and common-mode feedforward circuits, is added to ensure stability of the proposed filter. Measurement results show that the inter-modulation distortion of −40 dB can be achieved with 250 MHz 400 m Vpp balanced differential input signals. The filter works with a 1-V supply and its power consumption is 32 mW.

5.5.1 Introduction

As there is a great demand for lighter hand-held mobile phones and longer battery lifetime, low-voltage integrated circuit design solutions must be developed. Traditionally in CMOS technology, the baseband filters are realized with switched-capacitor techniques, but the SC filters are limited to low speed applications due to the sampling procedure and the high power requirement of the OPAMP. On the other hand, continuous-time G_m-C filter realizations can be easily implemented for the high speed applications and tend to be less power consuming. Furthermore, the filters do not require extra processing steps compared with the Active-RC structures, and their frequency tuning can be easily achieved. Thus, the $G_m - C$ filter appears to be a better candidate for the UWB wireless application, which uses pulse signals at a high speed.

For G_m-C filter implementation, the high performance transconductor would be the most important building block. In the design of transconductors, the transconductance should be tuned for compensating process tolerances and temperature variations without degrading the entire circuit performance. Besides, low supply voltage implementation should be adopted for a system-on-a-chip strategy. The performance of digital circuits does not degrade when using lower supply voltage. On the other hand, for analog circuits, the circuit performance is strongly affected by the low supply voltage. The linearity performance of transconductor will become worse at a low supply voltage and, hence a novel transconductor circuit design should be investigated.

In this section, the design of a high speed, low distortion and low supply voltage G_m-C filter for UWB wireless application is presented. The structure of high linearity transconductor with the pseudo-differential pair is discussed in Section 5.5.2. The common-mode control system with guaranteed stability is shown in Section 5.5.3. In Section 5.5.4, the design of fourth-order equiripple linear phase low-pass filter and the measurement results are discussed. The results are summarized in Section 5.5.5.

5.5.2 Proposed Operational Transconductance Amplifier

For the proposed transconductor, a pseudo-differential structure is used. The absence of the tail current source allows lower supply voltage and higher signal swing ranges. For large device length, the pseudo-differential pair with the source terminals connected to ground obtains a more linear performance in contrast with the fully differential architecture. However, in the technology down to smaller feature sizes, short channel effect occurs owing to the fact that the effective carrier mobility is a function of both the longitudinal and transverse electric fields, and thus the short channel effect and channel length modulation will degrade the linearity of the pseudo-differential structure.

Figure 5.16 shows the proposed transconductor circuit. The resistor R_{tune} is connected between current mirrors M3 and M4. By using the resistor, we can obtain a

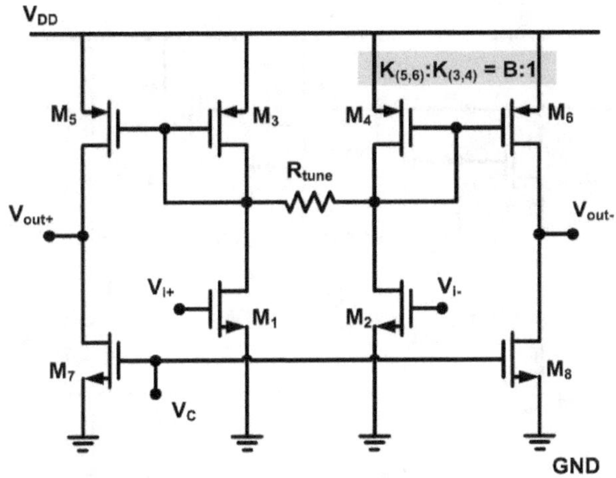

Fig. 5.16 Linearized low voltage transconductor circuit

voltage attenuation factor of $\zeta = gm_{(3,4)} \times R_{tune}/(2 + gm_{(3,4)} \times R_{tune})$ from the gate terminal to the drain terminal of transistors M1 and M2, where $1/gm_{(3,4)}$ is the output impedance of diode-connected transistors. If the value of $gm_{(3,4)} \times R_{tune}$ is small, a high attenuation factor can be achieved. In order to obtain the linearity performance, we adopt the usual mobility equation $\mu_{n,p} = \mu_0/(1 + \theta V_{ov})$, where μ_0 is the low-field mobility, θ is the mobility reduction coefficient, and V_{ov} is the MOS over-drive voltage. By taking the output current equation into a Taylor series expansion, the third-order harmonic distortion component of the transconductor can be expressed as

$$HD_3 \cong \frac{\theta}{16V_{ov(1,2)}(1 + \theta V_{ov(1,2)}\sqrt{\frac{BK_{(3,4)}}{K_{(1,2)}}})^2(2 + \theta V_{ov(1,2)}\sqrt{\frac{BK_{(3,4)}}{K_{(1,2)}}})} \times \sqrt{\frac{K_{(1,2)}}{BK_{(3,4)}}}\left(\frac{K_{(1,2)}}{K_{(3,4)}}\zeta V_p\right)^2 \tag{5.23}$$

where K_i is the device parameter of transistor M_i, V_p is the peak voltage of a sinusoid input signal, and B is the current mirror ration of M3 to M5 and M4 to M6.

The gain performance of the transconductor should be maintained as well. Usually, the gain of larger than 30 dB is sufficient for low Q low-pass filter design for correct signal transformation, and it could be maintained by designing a small aspect ratio of load transistors. Unfortunately, the use of R_{tune} would highly degrade the overall gain. The cascode circuit could be a possible solution, but it would not suitable for the low voltage design.

By taking the gain performance into consideration, the modified transconductor circuit is shown in Fig. 5.17. The transistor M11 has the same size as transistors M1 and M2, and thus it has the same drain current by giving the same input

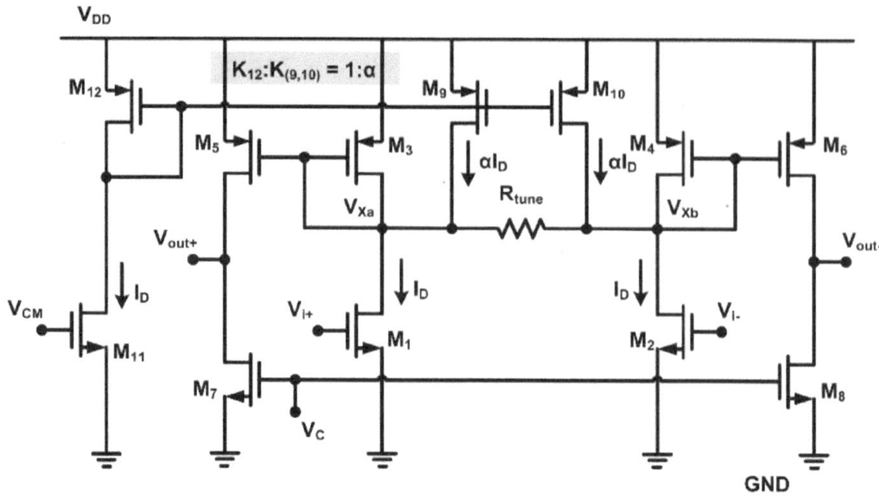

Fig. 5.17 The modified low voltage transconductor circuit

common-mode voltage at the gate terminal. Transistors M9 and M10 working in the saturation region with α times aspect ratio with respect to M12, are utilized to share part of DC drain current from M1 and M2. The increased gain can be expressed as

$$A_{enhanced} = \left(\frac{2 \times \sqrt{(1-\alpha)} + g_{m(3,4)} \times R_{tune}}{2 + g_{m(3,4)} \times R_{tune}} \right) \left(\frac{1}{1-\alpha} \right) \tag{5.24}$$

For example, if the value of α is 3/4, we can obtain that $1-\alpha$ would be equal to 1/4, and thus an enhanced gain in the range of 2–4 can be achieved. We should note that the structure also increases the linearity performance because the value of $gm_{(3,4)}$ in Fig. 5.17 has a factor of $\sqrt{(1-\alpha)}$ than the value of $gm_{(3,4)}$ in Fig. 5.16, and thus a smaller value of ζ can be obtained. In other words, it could relax the required sizes of transistors M3 and M4, and the value of R_{tune}. Moreover, transistors M9 and M10 would not affect the linearity performance owing to their high output impedance Therefore, the transconductor gain could be expressed as

$$\begin{aligned} A(s) = & \sqrt{\frac{K_{(1,2)}K_{(3,4)}}{1-\alpha}} \left(\frac{R_{\text{tune}}}{2 + \sqrt{K_{(1,2)}K_{(3,4)}(1-\alpha)}V_{ov(1,2)}R_{\text{tune}}} \right) \\ & \times \frac{1}{\lambda_{(5,6)}\left(1 + s\frac{C_L}{2\lambda_{(5,6)}B(1-\alpha)K_{(1,2)}V_{ov(1,2)}^2}\right)} \\ & \times \frac{1}{\left(1 + s\frac{R_{\text{tune}}C_X}{2+\sqrt{K_{(1,2)}K_{(3,4)}(1-\alpha)}V_{ov(1,2)}R_{\text{tune}}}\right)} \end{aligned} \tag{5.25}$$

where C_X and C_L are the capacitance at nodes $V_{X(a,b)}$ and $V_{out(+/-)}$, respectively. The term λ accounts for the effect of channel length modulation. The non-dominant pole should be carefully designed for smaller excess phase so that the transconductor can connect load capacitance C_L at the output to implement an integrator. The higher excess phase of the transconductor will cause the cutoff frequency of the resulting filter to deviate from desired value. We can see from (5.25) that the non-dominant pole would be affected by the value of α so as to contribute excess phase. Thus, the tradeoff among stability, gain, and linearity should be considered simultaneously.

5.5.3 Common-Mode Control Circuit

The differential-output transconductor requires proper common-mode control which not only stabilizes DC common-mode output voltage, but also suppresses the input common-mode signal at the output stage. For the fully-differential structure, the CMFB circuit is used as a negative feedback loop that fixes the output common-mode voltage. The tail current source in the fully-differential structure would restrict the input common-mode signal from variation. For the pseudo-differential structure, the way of suppressing the input common-mode signal is to include a CMFF circuit. Figure 5.18 illustrates an efficient common-mode control system, which is implemented by the CMFF circuit combined with the CMFB circuit. The CMFF circuit is composed by transistors MFFP1, MFFP2, and MFFN. The resistor R_{tune} is replaced by two MOS transistors operating in the linear region, and thus the output common-mode voltage can be simply sensed by next transconductor in the cascaded biquadratic structure. The value of R_{tune} is equal to $2/(K_{(r1,r2)} \times V_{ov(r1,r2)})$, where $V_{ov(r1,r2)}$ depends on V_{tune}. We should design the transistor with $K_{(5,6)} = K_{(FFP1,FFP2)}$, $2 \times K_{(7,8)} = K_{(FFN)}$, $K_{(1,2)} = K_{FBN}$, and $K_{(3,4)} = K_{FBP}$. In the figure, the CMFB loop is drawn in bold lines.

The CMFB circuit senses the output common-mode voltage from next transconductor and produces the corrected current to the CMFF circuit. As in the previous discussion, the common-mode control system is also designed under the consideration of high speed with stable phase margin. The open loop gain of the CMFB circuit is given by

$$A_{CMFB}(s) \cong g_{CMFB}(s) \times \frac{1}{(g_{o5,6} + g_{o7,8})} = \frac{g_{m(1,2)}}{(g_{o5,6} + g_{o7,8})(1-\alpha)} \times \frac{1}{\left(1 + s\frac{R_{tune}C_N}{2}\right)\left(1 + s\frac{C_C}{g_{mFFN}}\right)\left(1 + s\frac{C_L}{(g_{o5,6}+g_{o7,8})}\right)} \tag{5.26}$$

where C_C and C_N are the capacitances at nodes V_C and V_N, respectively. Again, the modified structure also increases the gain of the CMFB. The dominant pole of the circuit is $(g_{o5,6} + g_{o7,8})/C_L$ and the non-dominant high frequency poles are at g_{mFFN}/C_C and $2/R_{tune} \times C_N$. It is necessary to ensure that the frequency of the non-dominate poles would be larger than the unity-gain bandwidth.

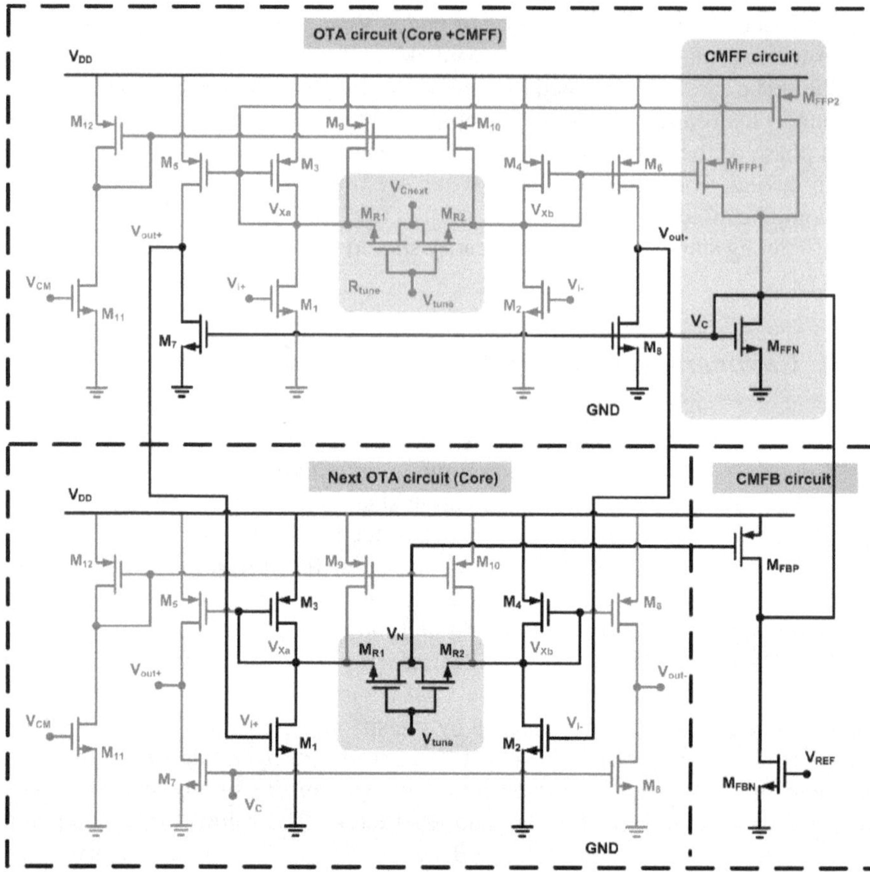

Fig. 5.18 The common-mode control system

The CMFF circuit uses replica current mirror of M5 and M6. It senses the input common-mode signal from two terminals of the resistor and cancels the common-mode signal variation. For the common-mode control operation, the error value will be sent from the CMFB part as an adaptive bias signal and combined with the CMFF part for overall stability. The input common-mode gain in our design can be also calculated, and the result of smaller than one at low frequency owing to the large value of g_{CMFB} can be obtained, so a higher CMRR can be expected.

5.5.4 Filter Implementation and Measurement Results

The fourth-order G_m-C low-pass filter based on the biquadratic section is designed. The requirement of the equiripple linear phase is due to the high speed pulse signal

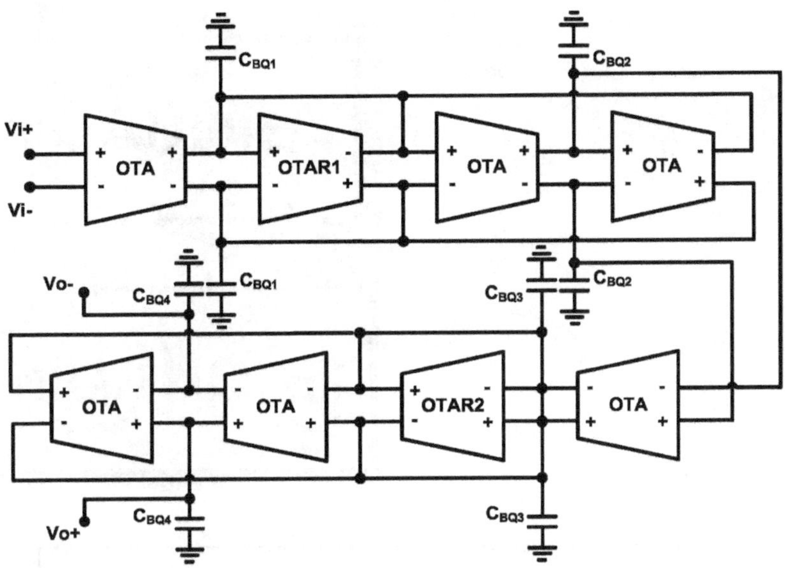

Fig. 5.19 The fourth-order equiripple linear phase filter

in the UWB wireless application to reduce the ISI effect. The implementation of the transconductor building block is shown in Fig. 5.19. The proposed transconductor circuit is used for all of the transconductor cells. The number of CMFB circuits is reduced due to the shared low impedance nodes of lossy integrators. The tuning strategy can be achieved by changing the voltage V_{tune} through the use of the MOS transistor operating in the linear region, and the maximum tuning voltage range of $V_{ov(1,2)} \times \sqrt{((1-\alpha)K_{(1,2)}/K_{(3,4)})}$ can be obtained.

The G_m-C filter is designed by using TSMC 180 nm CMOS process. The die photograph is shown in Fig. 5.20, where the active area is $2.4 \times 10^{-2}\,\text{mm}^2$. The parasitic capacitances and metal resistances are extracted to simulate the realistic environment. The transconductor circuit is simulated by connecting 1 pF capacitance at the output to work as an integrator. The gain of the transconductor is nearly 30 dB with 89° phase margin, and the unity-gain frequency is 250 MHz. The current mirror factor α is chosen to be 20/21 and the current mirror ratio B is set to be about 2. Figure 5.21 shows the measured magnitude response and the group delay of the filter. The magnitude response is normalized owing to attenuation from the output buffer. The -3 dB cutoff frequency of the low-pass G_m-C filter is 250 MHz and the CMRR is 35 dB. The group delay is almost constant over fc and the group delay ripple shows less than 2 ns over the range up to fc. The IM3 with two sinusoidal tones of 400 mV_{pp} is shown in Fig. 5.22. The IM3 is shown to be less than -40 dB at the speed of 250 MHz.

Fig. 5.20 Die micrograph

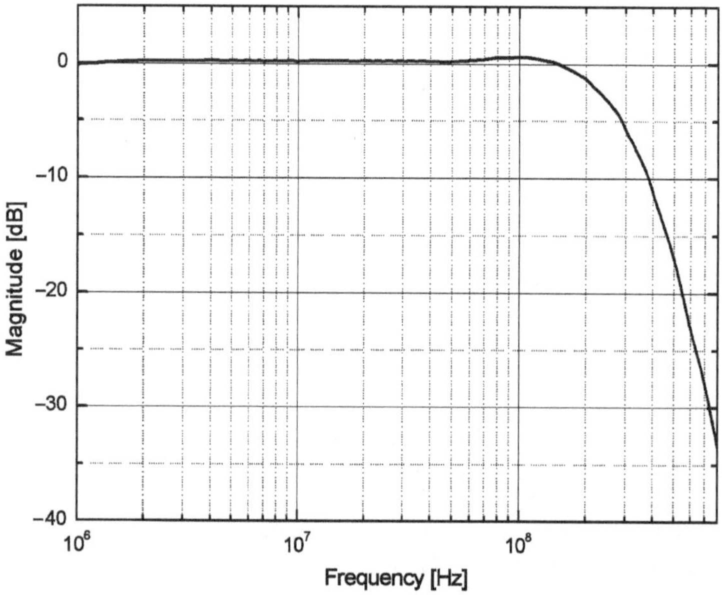

Fig. 5.21 The magnitude response and the group delay of the fourth-order equiripple linear phase G_m-C filter

5.5.5 *Summary*

A G_m-C filter implementation with an enhanced linearity CMOS transconductor has been presented, and the design is targeted for UWB wireless application. For the entire filter, the common-mode voltage of the pseudo-differential topology can be well defined by employing suitable CMFB and CMFF circuits. The circuit is

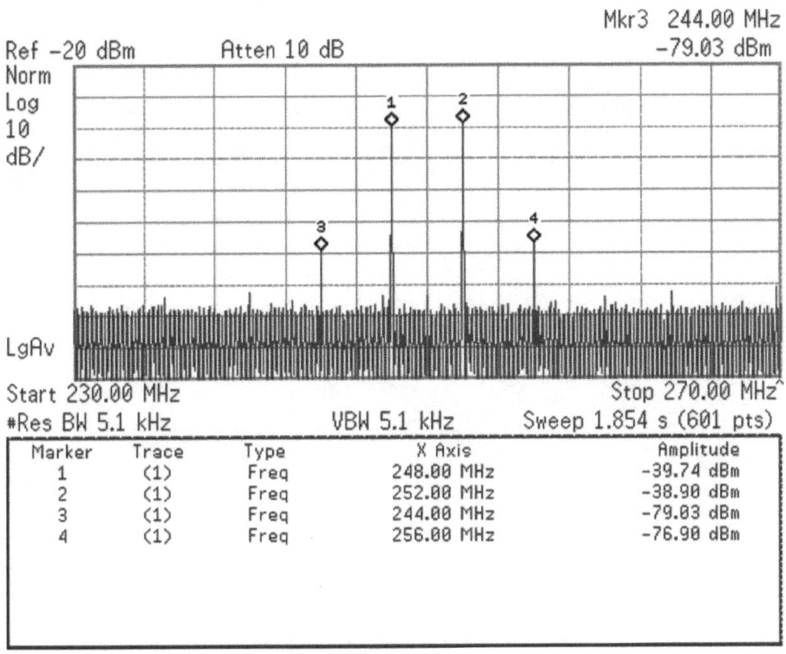

Fig. 5.22 Measured two tone inter-modulation distortion

Table 5.3 Comparison of previously reported works

Reference	[66] ESSCIRC 2006	[67] A-SSCC 2006	[68] ISSCC 2007	This work
Technology	0.12-μm CMOS	0.18-μm CMOS	0.13-μm CMOS	0.18-μm CMOS
Filter Order	3	8	5	4
-3dB Frequency	235MHz	250MHz	240MHz	250MHz
IM3/HD3	-43dB HD3 @ 50MHz	-37dB HD3 @ 150MHz	--	-40dB IM3 @ 250MHz
IIP3	--	--	-4.82dBV	3dBV
Supply Voltage	1.5	1.8	1.2	1

fabricated by the TSMC 180 nm CMOS process utilizing a 1-V supply with 32 mW power consumption. Measurement results show that high speed and excellent linearity. Table 5.3 summarizes the performance of this work with recently reported filters

Chapter 6
Conclusions

In this book, three transconductors are presented. The first transconductor is designed based on the cross-couple architecture under pseudo-differential structure, and the result shows that the −60 dB IM3 performance can be obtained at 40 MHz. The second transconductor is designed based on the cancellation to dominate the nonlinearity term while the common-mode signals are well controlled. For this circuit, the result shows that the −52 dB IM3 performance can be obtained at 50 MHz. The third transconductor is designed by using precise nano-scale process model, and the −65 dB HD3 can be obtained at 2 MHz for ADSL2 + application. A wide tuning rang filter is introduced. The transconductor is designed by using an equivalent active resistor, which combined with different transistor operation mode to extend the tuning range. The result shows a tuning ratio of 4,000. Three new multimode channel selection filters in the direct conversion receiver are presented for various wireless specifications. These designs are implemented based on the new building blocks through the third-order Butterworth filter. The first transconductor is designed based on the FVF structure through the use of an equivalent resistor. It can operate for IEEE 802.11 a/b/g/n Wireless LANs specifications. The other transconductors are implemented by using a linear voltage-to-current conversion as the input stage, and then a wide tuning multiplier is used as the output stage. One of the filters can be suitable for GSM/bluetooth/cdma2000/Wideband CDMA applications, and the other can be used for bluetooth/cdma2000/Wideband CDMA/IEEE 802.11 a/b/g/n Wireless LANs. These filters can also operate under low supply voltages. Two new high speed filters are presented. One is designed to meet the required performance of HDD read channel chip. The transconductor is designed by the inverter structure. A modified automatic tuning circuit is also proposed to compensate the deviation of cutoff frequency. The result shows the performance of 1 GHz cutoff frequency. The other high speed filter is designed to meet the specification of UWB. A high linearity transconductor based on the voltage attenuation and the gain enhancement structure is presented. We can obtain the measured cutoff frequency at 250 MHz with a −40 dB IM3 performance. The circuit constrains at 1-V supply are discussed, and the target of integrated system-on-a chip design can be achieved.

T.-Y. Lo and C.-C. Hung, *1V CMOS* G_m-C *Filters: Design and Applications*, Analog Circuits and Signal Processing,

References

1. R. H. Dennard, F. H. Gaensslen, H. N. Yu, V. L. Rideout, E. Bassous, and A. R. LeBlanc, "Design of Ion-Implanted MOSFET's With Very Small Physical Dimensions," *IEEE J. Solid-State Circuits*, vol. SC-9, pp. 256–268, Oct. 1974.
2. Y. A. El-Mansy, "On scaling MOS devices for VLSI," *IEEE Intl. Conf. on Circuits and Computers*, pp. 457–460, 1980.
3. T. Kuroda and T. Sakurai, "Overview of low-power {ULSI} circuit techniques," *IEICE Trans. on Electronics*, pp. 334–344, Apr. 1995.
4. J. E. Chung, M. C. Jeng, J. E. Moon, P. K. Ko, and C. Hu, "Performance and reliability design issue for deep-submicrometer MOSFET's," *IEEE Transactions on Electron Devices*, vol. 38, no. 3, pp. 545–554, Mar. 1991.
5. T. H. Ning, P. W. Cook, R. H. Dennard, C. M. Osburn, S. E. Schuster, and H. N. Yu, "1 um MOSFET VLSI technology: Part IV – hot-electron design constraints," *IEEE J. Solid-State Circuits*, vol. 14, no. 2, pp. 268–275, Apr. 1979.
6. C. Hu, S. C. Tam, F. C. Hsu, P. K. Ko, T. Y. Chan, and K. W. Terrill, "Hot-electron-induced MOSFET degradation-model, monitor, and improvement," *IEEE Transactions on Electron Devices*, vol. 35, no. 7, pp. 375–385, Feb. 1985.
7. R. W. Brodersen, "The network computer and its future," *IEEE Int. Solid-State Circuits Conf. Dig. Tech. Papers*, pp. 32–36, Feb. 1997.
8. H. Yasuda, "Multimedia impact on devices in the 21st century," *IEEE Int. Solid-State Circuits Conf. Dig. Tech. Papers*, pp. 28–31, Feb. 1997.
9. M. Ismail and T. Fiez, *Analog VLSI Signal and Information Processing*. New York: McGraw-Hill, 1994.
10. S. R. Zarabadi, M. Ismail, and C. C. Hung, "High performance analog VLSI computational circuits," *IEEE Journal of Solid-State Circuits*, vol. 33, no. 4, pp. 644–649, Apr. 1998.
11. T. Y. Lo, C. C. Hung, and M. Ismail, "A wide tuning range Gm-C filter for multimode direct-conversion wireless receivers," *Proc. IEEE Eur. Solid-State Circuits Conf.*, pp. 210–213, 2001.
12. T. Y. Lo and C. C. Hung, "A wide tuning range G_m-C continuous-time analog filter," *IEEE Trans. Circuits Syst. I, Reg. Papers.*, vol. 54, no. 4, pp. 713–722, Apr. 2007.
13. J. Galan, R. G. Carvajal, A. Torralba, F. Munoz, and J. Ramirez-Angulo, "A low-power low-voltage OTA-C sinusoidal oscillator with a large tuning range," *IEEE Trans. Circuits Syst. I, Reg. Papers.*, vol. 52, no. 2, pp. 283–291, Feb. 2005.
14. J. van Engelen, R. van de Plassche, E. Stikvoort, and A. Venes, "A sixth-order continuous-time bandpass sigma–delta modulator for digital radio IF," *IEEE J. Solid-State Circuits*, vol. 34, pp. 1753–1764, no. 12, Dec. 1999.
15. E. Sánchez-Sinencio and J. Silva-Martinez, "CMOS transconductance amplifiers, architectures and active filters: A tutorial," *Proc. IEEE Circuits Devices Syst.*, vol. 147, no. 1, pp. 3–12, Feb. 2000.

16. P. Pandey, J. Silva-Martinez, and X. Liu, "A CMOS 140-mW fourth-order continuous-time low-pass filter stabilized with a class AB common-mode feedback operating at 550 MHz," *IEEE Trans. Circuits Syst. I, Reg. Papers.*, vol. 53, no. 4, pp. 811–820, Apr. 2006.
17. M. Chen, J. Silva-Martinez, S. Rokhsaz, and M. Robinson, "A 2-V_{pp} 80-200-MHz fourth-order continuous-time linear phase filter with automatic frequency tuning," *IEEE Journal of Solid-State Circuits*, vol. 38, no. 10, pp. 1745–1749, Oct. 2003.
18. F. Rezzi, A. Baschirotto, and R. Castello, "A 3 V 12–55 MHz BiCMOS pseudo-differential continuous-time filter," *IEEE Trans. Circuits Syst. I*, vol. 42, no. 11, pp. 896–903, Nov. 1995.
19. A. E. Mourabit, G.-N. Lu, and P. Pittet, "Wide-linear-range subthreshold OTA for low-power, low-voltage, and low-frequency applications," *IEEE Trans. Circuits Syst. I, Reg. Papers.*, vol. 52, no. 8, pp. 1481–1488, Aug. 2005.
20. E. Rodriguez-Villegas, A. Yufera, and A. Rueda," A 1.25-V micropower Gm-C filter based on FGMOS transistors operating in weak inversion," *IEEE J. Solid-State Circuits*, vol. 39, no. 1, pp. 100–111, Jan. 2004.
21. A. J. Lopez-Martin, S. Baswa, J. Ramirez-Angulo, and R. G. Carvajal, "Low-voltage super class AB CMOS OTA cells with very high slew rate and power efficiency," *IEEE J. Solid-State Circuits*, vol. 40, no. 5, pp. 1068–1077, Mar. 2005.
22. A. Demosthenous and M. Panovic, "Low-voltage MOS linear transconductor/squarer and four-quadrant multiplier for analog VLSI," *IEEE Trans. Circuits Syst. I, Reg. Papers.*, vol. 52, no. 9, pp. 1721–1731, Sep. 2005.
23. P. Kallam, E. Sánchez-Sinencio and A. Karsilayan, "An enhanced adaptive Q-tuning scheme for a 100 MHz fully symmetric OTA-based bandpass filter," *IEEE Journal of Solid-State Circuits*, vol. 38, no. 4, pp. 585–593, Apr. 2003.
24. T. Y. Lo and C. C. Hung, "A high speed and high linearity OTA in 1-V power supply voltage," *Proc. ISCAS*, pp. 1864–1867, 2006.
25. U. Yodprasit and C. C. Enz, "A 1.5-V 75-dB dynamic range third-order Gm-C filter integrated in a 0.18-μm standard digital CMOS process," *IEEE Journal of Solid-State Circuits*, vol. 38, no. 7, pp. 1189–1197, Jul. 2003.
26. A. Nader Mohieldin, E. Sánchez-Sinencio, and J. Silva-Martínez, "A fully balanced pseudo-differential OTA with common-mode Feedforward and inherent common-mode feedback detector," *IEEE Journal of Solid-State Circuits*, vol. 38, no. 4, pp. 663–668, Apr. 2003.
27. I. S. Han, "A novel tunable transconductance amplifier based on voltage-controlled resistance by MOS Transistors," *IEEE Trans. Circuits Syst. II, Expr. Briefs.*, vol. 53, no. 8, pp. 662–666, Aug. 2006.
28. A. Lewinski, and J. Silva-Martinez, "OTA linearity enhancement technique for high frequency applications with IM3 below -65 dB," *IEEE Trans. Circuits Syst. II*, vol. 51, no. 10, pp. 542–548, Oct. 2004.
29. A. J. Lopez-Martin, J. Ramirez-Angulo, C. Durbha, and R. G. Carcajal, "A CMOS transconductor with multidecade tuning using balanced current scaling in moderate inversion," *IEEE Journal of Solid-State Circuits*, vol. 40, no. 5, pp. 1078–1083, May 2005.
30. Y. S. Youn, J. H. Chang, K. J. Koh, Y. J. Lee, H. K. Yu, "A 2 GHz 16 dBm IIP3 low noise amplifier in $0.25\,\mu$m CMOS technology," *IEEE Int. Solid-State Circuits Conf. Dig. Tech. Papers*, pp. 452–453, Feb. 2003.
31. C. Fager, J. C. Pedro, N. B. Carvalho, H. Zirath, F. Fortes, and M. J. Rosário," A comprehensive analysis of IMD behavior in RF CMOS power amplifiers," *IEEE Journal of Solid-State Circuits*, vol. 39, no. 1, pp. 24–34, Jan. 2004.
32. S. H. Tang, K. H. Kim, Y. H. Kim, Y. You, and K. R. Cho, "A novel CMOS operational transconductance amplifier based on a mobility compensation technique," *IEEE Trans. Circuits Syst. II, Expr. Briefs*, vol. 52, no. 1, pp. 37–42, Jun. 2005.
33. X. Fan and P. K. Chan, "Analysis and design of low-distortion CMOS source followers," *IEEE Trans. Circuits Syst. I*, vol. 52, no. 8, pp. 1489–1501, Aug. 2005.
34. F. Behbahani, T. Weeguan, A. Karimi-Sanjaani, A. Roithmeier, and A. A. Abidi, "A broadband tunable CMOS channel-select filter for a low-IF wireless receiver," *IEEE J. Solid-State Circuits*, vol. 35, no. 4, pp. 476–489, Apr. 2000.

35. I. S. Han, "A novel tunable transconductance amplifier based on voltage-controlled resistance by MOS transistors," *IEEE Trans. Circuits Syst. II, Expr. Briefs*, vol. 53, no. 8, pp. 662–666, Aug. 2006.
36. F. A. P. Baruqui and A. Petraglia, "Linearly tunable CMOS OTA with constant dynamic range using source-degenerated current mirrors," *IEEE Trans. Circuits Syst. II, Expr. Briefs*, vol. 53, no. 9, pp. 797–801, Sep. 2006.
37. S. Pavan and Y. Tsividis, *High Frequency Continuous Time Filters in Digital CMOS Process*, Norwell, Massachusetts: Kluwer, 2000.
38. S. Chatterjee, Y. Tsividis, and P. Kinget, "0.5-V analog circuit techniques and their application in OTA and filter design," *IEEE J. Solid-State Circuits*, vol. 40, no. 12, pp. 2373–2387, Dec. 2005.
39. K. C. Kuo and A. Leuciuc, "A linear MOS transconductor using source degeneration and adaptive biasing," *IEEE Trans. Circuits Syst. II*, vol. 48, no. 10, pp. 937–943, Oct. 2001.
40. B. Razavi, *Design of Analog CMOS Integrated Circuit*, New York: McGraw-Hill, 2001.
41. M. Steyaert, J. Silva-Martinez and W. Sansen, "High-frequency saturated CMOS floating resistor for fully-differential analogue signal processors," *IEEE Electronics Letters*, pp. 1609–1611, Aug. 1991.
42. L. P. Huelsman, *Active and Passive Analog Filter Design*, New York: McGraw-Hill, 1993.
43. C. C. Hung, K. A. Halonen, M. Ismail, V. Porra and A. Hyogo, "A low-voltage, low-power CMOS fifth-order elliptic GM-C filter for baseband mobile, wireless communication," *IEEE Trans. Circuits Syst. Video Technol.*, vol. 7, pp. 584–593, Aug. 1997.
44. D. Chamla, A. Kaiser, A. Cathelin, and D. Belot, "A G_m-C low-pass filter for zero-IF mobile applications with a very wide tuning range," *IEEE J. Solid-State Circuits*, vol. 40, no. 7, pp. 1143–1450, Jul. 2005.
45. G. Bollati, S. Marchese, M. Demicheli, and R. Castello, "An eighth-order CMOS low-pass filter with 30–120 MHz tuning range and programmable boost," *IEEE J. Solid-State Circuits*, vol. 36, no. 7, pp. 1056–1066, Jul. 2001.
46. J. A. De Lima and C. Dualibe, "A linearly tunable low voltage CMOS transconductor with improved common-mode stability and its application to *Gm-C* filters," *IEEE Trans. Circuits Syst. II*, vol. 48, no. 7, pp. 649–660, Jul. 2001.
47. S. Pavan, Y. P. Tsividis, and K. Nagaraj, "Widely programmable high frequency continuous-time filters in digital CMOS technology," *IEEE J. Solid-State Circuits*, vol. 35, no. 4, pp. 503–511, Apr. 2000.
48. N. Rao, V. Balan, and R. Contreras, "A 3-V 10–100 MHz continuous-time seventh-order 0.05 equiripple linear phase filter," *IEEE J. Solid-State Circuits*, vol. 34, pp. 1676–1682, Nov. 1999.
49. R. G. Carvajal, J. Ramirez-Angulo, A. J. Lopez-Martin, A. Torralba, J. A. G. Galan, A. Carlosena, and F. M. Chavero, "The Flipped Voltage Follower: A Useful Cell for Low-Voltage Low-Power Circuit Design," *IEEE Trans. Circuits. Syst. I, Reg. Papers.*, vol. 52, no. 7, pp. 1276–1291, Jul. 2005.
50. R. Alini, A. Baschirotto, and R. Castello, "Tunable BiCMOS continuous-time filter for high-frequency applications," *IEEE J. Solid-State Circuits*, vol. 27, no. 12, pp. 1905–1915, Dec. 1992.
51. D. A. Johns and K. Martin, *Analog Integrated Circuit Design*, New York: Wiley, 1997.
52. B. Fotouhi, "ALL-MOS voltage-to-current converter," *IEEE J. Solid-State Circuits*, vol. 36, no. 1, pp. 147–151, Jan. 2001.
53. The BSIM model, BSIM Research Group at UC Berkeley. Available: http://www-device.eecs.berkeley.edu/~bsim3/bsim4.html
54. S. Hori, T. Maeda, H. Yano, N. Matsuno, K. Numata, N. Yoshida, Y. Takahashi, T. Yamase, R. Walkington, and H. Hida, "A widely tunable CMOS Gm-C filter with a negative source degeneration resistor transconductor," *Proc. IEEE Eur. Solid-State Circuits Conf.*, pp. 449–452, 2003.
55. S. Hori, T. Maeda, N. Matsuno, and H. Hida, "Low-power widely tunable Gm-C filter with an adaptive DC-blocking, triode-biased MOSFET transconductor," *Proc. IEEE Eur. Solid-State Circuits Conf.*, pp. 99–102, 2004.

56. S. D'Amico, V. Giannini, and A. Baschirotto, "A 4th-order active-Gm-RC reconfigurable (UMTS/WLAN) filter," *IEEE J. Solid-State Circuits*, vol. 40, no. 7, pp. 1143–1450, Jul. 2006.
57. L. T. Bruton, *RC-Active Circuits, Theory and Design*, Englewood Cliffs, New Jersey: Prentice Hall, 1980.
58. N. Rao, V. Balan, R. Contreras, J. Chen, and Y. Wang, "A 150 MHz continuous-time 7th-order 0.05° equiripple linear phase filter," *Proc. IEEE ISCAS*, pp. 664–667, 1999.
59. Y. Tsividis and J. Voorman, *Integrated Continuous-Time Filters: Principles, Design and Applications*, New York: IEEE Press, 1992.
60. B. Nauta, "A CMOS transconductance-C filter technique for very high frequencies," *IEEE J. Solid-State Circuits*, vol. 27, no. 2, pp. 142–153, Feb. 1992.
61. Y. Oowaki, M. Noguchi, S. Takagi, D. Takashima, M. Ono, Y. Matsunaga, K. Sunouchi, H. Kawaguchiya, S. Matsuoka, M. Kamoshida, T. Fuse, S. Watanabe, A. Toriumi, S. Manabe, and A. Hojo, "A sub-0.1 µm circuit design with substrate-over-biasing," *IEEE Int. Solid-State Circuits Conf. Dig. Tech. Papers*, pp. 88–89, 1998.
62. M. Miyazaki, G. Ono, and K. Ishibash, "A 1.2-GIPS/W microprocessor using speed-adaptive threshold-voltage CMOS with forward bias," *IEEE J. Solid-State Circuits*, vol. 37, no. 2, pp. 142–153, Feb. 2002.
63. J. Silva-Martinez, Joseph Adut, Jose Miguel Rocha-Perez, Moises Robinson, and Shahriar Rokhsaz, "A 60 mW, 200 MHz continuous-time seventh-order linear phase filter with on-chip automatic tuning system," *IEEE J. Solid-State Circuits*, vol. 38, no. 2, pp. 216–225, Feb. 2003.
64. P. Pandey, J. Silva-Martinez, and X. Liu, "A CMOS 140-mW fourth-order continuous-time low-pass filter stabilized with a class AB common-mode feedback operating at 550 MHz," *IEEE Trans. Circuits Syst. I, Reg. Papers.*, vol. 53, no. 4, pp. 811–820, Apr. 2006.
65. S. Dosho, T. Morie, and H. Fujiyama, "A 200 Mhz seventh-order equiripple continuous-time filter by design of nonlinearity suppression in 0.25-µm CMOS process," *IEEE J. Solid-State Circuits*, no. 5, pp. 559–565, May 2002.
66. R. Kolm and H. Zimmermann, "A 3rd-Order 235 MHz low-pass gmC-filter in 120 nm CMOS," *Proc. IEEE Eur. Solid-State Circuits Conf.*, pp. 215–218, 2006.
67. C. C. Lee and T. Y. Yang, "A tuning technique for bandwidth of programmable gain filter," *IEEE Asia Solid-State Circuits Conf.*, pp. 175–178, 2006.
68. V. Sarri, M. Kaltiokallio, S. Lindfors, J. Ryynänen, K. Halonen, "A 1.2 V 240 MHz CMOS continuous-time low-pass filter for a UWB radio receiver," *IEEE Int. Solid-State Circuits Conf. Dig. Tech. Papers*, pp. 122–123, 2007.

About the Authors

Tien-Yu Lo received the B.S., M.S., and Ph.D. degrees in Communication Engineering from National Chiao Tung University, Taiwan, in 2001, 2003 and 2007, respectively. Since 2008, he has been with the Analog Circuit Design Division, MediaTek Inc., Taiwan, as an analog IC designer. His research interests include analog and mixed-signal circuit design with particular focus on the topic of continuous-time filters and analog-to-digital converters.

Chung-Chih Hung received the B.S. degree in Electrical Engineering from National Taiwan University, Taipei, Taiwan, in 1989, and the M.S. and Ph.D. degrees in electrical engineering from The Ohio State University, Columbus, Ohio, USA, in 1993 and 1997, respectively. From 1989 to 1991, he served in the Taiwan Marine Corps as a communication officer. From 1997 to 2003, he worked for several IC design companies in San Jose, CA, and San Diego, CA, where he held analog circuit design manager and director positions. Since 2003, he has been with National Chiao Tung University, Taiwan, where he is currently an Associate Professor in the Department of Electrical Engineering. His research interests include the design of analog and mixed-signal integrated circuits for communication and high-speed applications.

FSC
www.fsc.org
MIX
Papier | Fördert
gute Waldnutzung
FSC® C083411

Kolibri 360 GmbH
Ferdinand-Jühlke-Straße 7
99095 Erfurt, Deutschland
produktsicherheit@kolibri360.de